CONTENTS

of Transportation Expressions, produced and maintained by the Bureau of Transportation Statistics, US Department of Transportation (1996). With assistance from Dr. Kumares Sinha of Purdue University, the US Department of Transportation granted T&DI/ASCE permission to translate and publish *Transportation Expressions* into the Chinese language. Adhering to the recommendation of the Expert Panel to focus on one or two subject areas as a first effort, this publication is a collection of highway and railway expressions contributed to the BTS Glossary by four industry and governmental organizations: the American Trucking Association (ATA), the US Federal Highway Administration (FHWA), the US Federal Railway Administration (FRA) and the National Highway Transporation Safety Administration (NHTSA).

Under the supervision of several academic scholars and faculty members, students enrolled in graduate transportation programs in China and the US completed the initial translation of *Transportation Expressions* from English to Chinese. The translations were edited by the group supervisors and submitted for review and approval by NACOTA volunteers. Three or more experts reviewed each section, and revisions were synthesized and forwarded to the Quality Assurance and Qualify Control (QAQC) Committee.

Led by Dr. Xiaoduan Sun, a transportation expert with extensive experience both in China and in the US, the QAQC Committee was comprised of three NACOTA members, later expanded to four. Each member of the Committee had extensive transportation experience in both Chinese and English language speaking environments. The QAQC Committee reviewed the glossary terms to ensure that the translations not only reflected the original meaning of the English terms but that they also followed native Chinese semantics. ASCE Publications staff handled the formatting so that the text would appear in a clear and easy-to-read manner. Jonathan Esslinger, Director of T&DI, provided institutional support, guiadance and encouragement throughout the process.

It is our hope that this first volume of the T&DI/ASCE – NACOTA Transportation Glossary will be useful to the many transportation professionals in both China and the West who seek better interaction and engagement in highway and railway subject areas, across English-Chinese language barriers. We thank the leaders and members of T&DI and NACOTA for their support and encouragement of this endeavor.

Rongfang (Rachel) Liu, 刘荣芳, and

Eva Lerner-Lam, 林意华

EXPERT REVIEW PANEL

Fang, Shouen
General Secretary, Professor,
College of Transportation,
Tongji University, Shanghai

方守恩
书记，教授
交通学院
同济大学

Gao, Li
Chair,
Department of Transportation Engineering,
Beijing Institute of Technology.

高利
系主任
交通工程系
北京理工大学

Guan, Hongzhi
Professor,
Transportation Research Center, Beijing
University of Technology.

关宏志
教授
交通研究中心
北京工业大学

Guo, Zhongyin
Dean,
College of Transportation,
Tongji University,

郭中印
院长
交通学院
同济大学

Hong, Yan
Deputy Director and Professor,
Institute of Research.
Ministry of Railways

洪艳
付主任，研究员
铁路研究所
铁道部

Lu, Huapu
Director,
Transportation Research Center,
Tsinghua University

陆化普
主任
交通研究中心
清华大学

Shao, Chunfu
Dean,
College of Transportation,
Beijing Transportation University

绍春福
院长
交通学院
北京交通大学

Tang, Chengcheng
Deputy Director,
Highway Research Institute,
Ministry of Construction

唐铠铠
副主任
高速路研究所
建设部

Wang, Wei
Dean,
College of Transportation,
South East Transportation University

王炜
院长
交通学院
东南大学

Wang, Yuanqing
Chair,
Department of Transportation,
Chang An University

王元庆
系主任
交通系
长安大学

Yan, Xinping
Vice President for Academic Affairs,
Wuhan University of Technology.

严新平
付校长
武汉理工大学

Yang, Kaizhong
Dean,
College of Government,
Beijing University

杨开忠
院长
政府学院
北京大学

Zhang, Jin
Dean,
College of Logistic,
Southwest Transportation University

张锦
院长
物流学院
西南交大

Zhou, Wei
Director,
Institute of Transportation Research,
Department of Transportation

周伟
所长
交通研究所
交通部

TRANSLATORS

Chen, Xuemei	陈雪梅	Wang, Dong	王东
Deng, Yi	邓怡	Wang, Yindu	王尹都
Guan, Changqian	关常谦	Wu, Shaobin	吴绍斌
Guo, Shanshan	郭姗姗	Xiong, Hui	熊辉
Hao, Wei	郝威	Xu, Tianci	许天赐
He, Yang	何杨	Yang, Fei	杨菲
Hu, Yaxin	胡亚馨	Yang, Fei (M)	杨飞
Li, Guilin	李桂林	Yao, Guozhong	姚国仲
Liu, Xiangmei	刘香梅	Zhao, Yanan	赵亚男
Tan, Huachun	谭华春	Zheng, Zuduo	郑祖舵

REVIEWERS

Chen, Cynthia	陈秋子	Wang, Jun	王军
Chu, Lianyu	初连禹	Wang, Linbing	汪林兵
Dai, Songtao	戴松涛	Wang, Xiubin	王秀斌
Fang, Fang	方芳	Wang, Xuesong	王雪松
Fang, Yingwu	房英武	Wang, Yinhai	王银海
Guo, Hua	郭骅	Wang, Yunshi	王云是
Huang, Allen	黄建华	Xia, Jingxin	夏井新
Huo Hong	霍洪	Xiong, Cherry	熊谦
Jin, Xia	金霞	Xu, Lei	许磊
Li, Qiang	李强	Yang, Jiawen	杨佳文
Liu, Chiu	刘遒	Yang, Qi	杨齐
Liu, Feng	刘枫	Yang, Qingyan	杨庆岩
Liu, Jenny	刘娟玉	Yi, Ping	易平
Liu, Xiaobo	刘晓波	You, Zhanping	尤占平
Pei, Jianping	裴剑平	Yu, Xiong	余雄
Tao, Reihua	陶瑞华	Zhang Jian	张健
Shi, Xianming	石鲜明	Zhang, Ming	张明

Sun, Dazhi	孙大志	Zhang, Zhongjie	张忠捷
Tao, Zongwei	陶宗伟	Zhang, Yujiang	张玉江
Tian, Zongzhong	田宗忠	Zhong , Ming	钟明
Wang, Alan	王键雄	Zhou, Huaguo	周华国

QUALITY ASSURANCE/QUALITY CONTROL (QAQC) COMMITTEE

Deng, Yi	邓怡	Sun, Xiaoduan	孙小端
Lu, Jian	陆键	Yu, Lei	于磊

ACRONYMS

English	Acronym	Chinese
Aerospace Industries Association	AIA	航空工业协会
Air Transport Association	ATAB	空中运输协会
Airports Council International	ACI	国际机场委员会
American Gas Association	AGA	美国汽油协会
American Public Transit Association	APTA	美国公共运输协会
American Trucking Associations	ATA	美国卡车协会
Association of American Railroads	AAR	美国铁路协会
Bureau of Economic Analysis	BEA	经济分析局
Bureau of Transportation Statistics	BTS	交通统计局
Bureau of the Census	BOC	人口普查局
Code of Federal Regulations	CFR	联邦规章代码
Department of Army, Army Corps of Engineers	DoD/COE	陆军部, 陆军工程师团
Department of Army, Military Traffic Management Command	MTMC	陆军部, 军用交通命令管理
Department of Energy	DOE	能源部
Eno Transportation Foundation, Inc.	ENO	ENO 交通基金会
Federal Aviation Administration	FAA	联邦飞行管理局
Federal Highway Administration	FHWA	联邦高速公路管理局
Federal Railroad Administration	FRA	联邦铁路管理局
Federal Transit Administration	FTA	联邦公交管理局
General Services Administration	GSA	普通服务管理局
Maritime Administration	MARAD	海运管理局
Metro Magazine	MM	地铁杂志
Missoula Office of Community Development	MOCD	密苏拉社区发展办公室

English	Acronym	Chinese
National Highway Traffic Safety Administration	NHTSA	国家高速交通安全管理局
National Research Council. Transportation Research Board	TRB	国家研究委员会,交通研究部
National Safety Council	NSC	国家安全部
National Transportation Safety Board	NTSB	国家交通安全部
Office of the Federal Register	OFR	联邦注册办公室
Office of the Secretary	OST	秘书办公室
Research and Special Programs Administration	RSPA	研究及专门项目管理部
Saint Lawrence Seaway Development Corporation	SLSDC	圣劳伦斯海上航道发展公司
Tennessee Department of Transportation	TNDOT	田纳西交通部
Texas Dept. of Transportation	TXDOT	德州交通部
Truck Index, Inc.	TII	卡车指数公司
U.S. Travel and Tourism Administration	USTTA	美国旅行观光局
UN/EUROSTAT/ECMT	UN	英国/欧洲统计
United States Coast Guard	USCG	美国海岸看守
United States Department of the Interior	DOI	美国内务部
Virginia Dept. of Transportation	VDOT	弗及尼亚州交通部

English	Chinese
Abbreviated Injury Scale: An integer scale developed by the Association for the Advancement of Automotive Medicine to rate the severity of individual injuries.	**简要损伤等级:** 由汽车医药发展协会提出的, 用于度量个人受伤严重性的整数指标.
Acceleration Power: Measured in kilowatts. Pulse power obtainable from a battery used to accelerate a vehicle. This is based on a constant current pulse for 30 seconds at no less than 2/3 of the maximum open-circuit-voltage, at 80% depth-of-discharge relative to the battery's rated capacity and at 20 degrees Celsius ambient temperature.	**加速度功率:** 用千瓦来度量. 从车辆的电瓶 获得的脉冲功率。这是基于环境温度 20 摄氏度，恒流脉冲为 30 秒, 在不低于 2/3 的最高开路电压，相对电池的额定容量 80%深度的放电时的取值。
Access Restrictions Road Gate: Constraints on use.	**限制出入通道闸门:** 限制道路的使用.
Access Rights: This element identifies who has acquired legal access rights over a road segment.	**驶入权:** 即驶入某一路段所需具备的法律权利.
Accessory or Auxiliary Equipment: A particular item of equipment added to a vehicle to aid or contribute to the vehicle's operation and/or mission.	**配件或辅助设备:**附加于车辆上的，用与帮助车辆运行和/或完成任务的特殊设备.
Accident: An occurrence involving a commercial motor vehicle operating on a public road which results in one of the following: a fatality; bodily injury to a person, who as a result of the injury, immediately receives medical treatment away from the scene of the accident; or one or more motor vehicles incurring disabling damage as a result of the accident, requiring the vehicle to be transported away from the scene by a tow	**事故:** 在公共道路上发生的和商业运营机动车有关的事件, 并造成以下结果之一:死亡;身体伤害,伤员被立即撤离事故现场并接受医治;或者由于事故，造成一辆或多辆汽车损坏,需要将车拖离现场.它不包括：那些只涉及从静止机动车上下乘客时发生的事件;那些只涉及装卸货物时发生的事件;以及营运小

English	Chinese

truck or other vehicle. It does not include: an occurrence involving only boarding and alighting from a stationary motor vehicle; an occurrence involving only the loading or unloading of cargo; or an occurrence in the course of the operation of a passenger car or a multipurpose passenger vehicle, as defined in 49 CFR 571.3, by a motor carrier and is not transporting passengers for hire or hazardous materials of a type and quantity that require the motor vehicle to be marked or placarded in accordance with 49 CFR 177.823.

轿车或营运多功能客车（具体见未来研究中心 CFR 571.3 第 49 篇中的定义）在没有以赢利为目的的情况下运送乘客和没有运送 49 CFR 177.823 规定的化学制品的情况下，运行途中发生的事故。

Accident: 1) An event that involves the release of gas from a pipeline or of liquefied natural gas or gas from an LNG facility resulting in a death, or personal injury necessitating in-patient hospitalization; or estimated property damage, including cost of gas lost, of the operator or others, or both, of $50,000 or more; 2) an event that results in an emergency shutdown of an LNG facility; 3) an event that is significant, in the judgment of the operator, even though it did not meet the criteria of 1) or 2).

事故: 1) 事件所涉及的管道泄漏的气体、液化天然气或液化天然气设施泄漏的气体造成人员死亡，或致使受伤人员必须接受住院治疗；估计财产损失（其中包括泄漏的气体对于经营者和其他方中之一或双方的成本）在 5 万美元以上;2)导致液化天然气设施紧急关闭的事件;3)即使不符合 1)或 2) 的标准，但是经营者经判断认为严重的事件。

Accident: Occurrence in a sequence of events that produces unintended injury, death or property damage. Accident refers to the event, not the result of the event.

意外事故: 发生的一连串事件,造成意外伤亡或财物损失. 意外,指的是事件本身,而不是事件的结果.

Accident Classes: Used to categorize commercial vehicle accidents according to accident severity (i.e., fatal accidents, injury

事故类别: 根据意外事故的严重程度来给商用车事故分类(即死亡事故、受伤事故和物损

English	Chinese
accidents, and property damage accidents).	事故).
Accident Consequences: The physical results of motor vehicle accidents. Consequences include fatalities, injuries, and property damage.	**事故后果:** 机动车事故造成的后果；包括: 死亡、受伤和财产损失.
Accident Severity: Measures the seriousness of an accident according to the type and quantity of the accident's consequences. Fatalities are more severe than injuries, and injuries are more severe than property damage.	**事故的严重程度:** 根据事故后果的类型和大小来衡量该事故的严重程度。死亡比受伤严重, 受伤比财产损失严重.。
Accident Type: An accident type is classified as either "collision" or "non-collision."	**事故类型**：可分为"碰撞事故"和"非碰撞事故"。
Actual Severity: On scene evaluation of the degree of danger that existed. An "after-the-fact" evaluation by the reporting unit.	**实际的严重程度**：在现场评估的存在危险的程度. 报告部门给出的"事后"评价.
Ad Valorem: A charge levied on persons or organizations based on the value of transaction. It is normally a given percentage of the price at the retail or manufacturing stage and is a common form of sales tax; e.g. Federal excise tax on new trucks and trailers.	**按价计税**：基于商品交易值向个人或机构征收的费用。它通常是在销售或制造阶段价格的某一百分比，是消费税的一种常见形式（如: 联邦政府对新购卡车和拖车征收的消费税）。
Additional Vehicle: A vehicle added to the inventory of a Fleet Management Center to fill a new program or to expand on an existing program of a participating agency.	**新增车辆:** 增加到车队管理中心中的新车辆, 可以用来完成新的任务或拓展现有任务.
Adjustable Speed Drives: Drives that save energy by ensuring the motor's speed is	**调速驱动**：通过使机动车的速度与车辆的负荷相匹配来实现

English	Chinese
properly matched to the load placed on the motor. Terms used to describe this category include polyphase motors, motor oversizing, and motor rewinding.	节约能源驾驶. 形容这一类的词语包括多相电机, 超大型电机,电机复卷.
Administrative Road: Consists of all public and non-public roads intended to be used principally for administrative purposes. It includes roads servicing employee residential areas, maintenance areas and other administrative developments, as well as restricted patrol roads, truck trails, and similar service roads.	**行政道路**：包括所有公共和非公共道路, 主要目的是用于行政.它包括连接职工住宅区,维修地区, 和其他行政方面的道路,以及限制公众使用的巡逻路,卡车通道,和类似的服务性道路.
Adverse Weather: The weather conditions considered by the operator in identifying the response systems and equipment to be deployed in accordance with a response plan, including wave height, ice, temperature, visibility, and currents within the inland or Coastal Response Zone (defined in the National Contingency Plan (40 CFR 300)) in which those systems or equipment are intended to function.	**恶劣天气**：操作人员所考虑的天气情况, 以便根据应急预案来确定反应系统和设备, 其中包括在应急系统和设备适用的内陆或沿海回应区的波高,冰,温度, 能见度和洋流.
Agricultural Commodity Trailer: A trailer that is designed to transport bulk agricultural commodities in off-road harvesting sites and to a processing plant or storage location, as evidenced by skeletal construction that accommodates harvest containers, a maximum length of 28 feet, and an arrangement of air control lines and reservoirs that minimizes damage in field operations.	**农业商品拖车**: 该车辆将大宗农产品从远离道路的收获地点运送到加工厂或存储地点。 这些车辆的特点是车体宽大适用于集装箱,最大长度为 28 英尺并安排空中控制线和框架结构,最大限度减少在实地作业的损失.

English	Chinese
Air Brake: A brake in which the mechanism is actuated by manipulation of air pressure. The term is often used to describe brakes that employ air under pressure above atmospheric, in contrast to vacuum brakes, which employ pressure below atmospheric.	**空气制动**: 通过操纵空气压力启动制动机制。 这一术语经常被用来形容运用空气压力高于大气压的方法进行刹车。与低于气压的真空刹车正好相反
Alcohol Concentration (AC): The concentration of alcohol in a person's blood or breath. When expressed as a percentage it means grams of alcohol per 100 milliliters of blood or grams of alcohol per 210 liters of breath.	**酒精浓度 (AC)**: 一个人的血液或呼吸中的酒精含量。用百分比表示时相当于每 100 毫升血液中或每 210 升呼气中所含的酒精克数
Alcohol Involvement: A fatality or fatal crash as alcohol-related or alcohol involved if 1) Either a driver or a nonmotorist (usually a pedestrian) had a measurable or estimated blood alcohol concentration (BAC) of 0.01 grams per deciliter (g/dl) or above. Probabilities of alcohol involvement are now calculated for each driver, pedestrian, or crash. 2) Coded by police when evidence of alcohol is present. This code does not necessarily mean that a driver, passenger or nonoccupant was tested for alcohol.	**酒精的参与** : 与酒精相关的或酒精参与的死亡或死亡事故,满足以下条件 1)司机或者非机动车用路者(通常为行人)中的任一方具有可以测量或估计出的血液中酒精浓度(BAC)等于或超过 0.01 克/分升（g/dl）。目前针对每个司机,行人或事故都要进行有关酒精参与程度的检测。2) 当有酒精参与的证据时,警方会进行编码记录.编码并不一定意味着机动车驾驶员,乘客或者其他用路者被测试了酒精含量。
Alternative Fuel Capacity: The on-site availability of apparatus to burn fuels other than natural gas.	**替代燃料能力** : 机器的具有燃烧除天然气以外的其他燃料的能力.
Ambulance or Rescue Service: Establishments primarily engaged in	**救护车或救助服务机构** : 这类的机构主要从事救护车或救援

English	Chinese
furnishing ambulances or rescue services, except by air, where such operations are primarily within a single municipality, contiguous municipalities, or a single municipality and its suburban areas.	服务（航空方式除外），主要是以一个市，毗连的市镇，或一个市及其郊区为服务范围。
Amtrak: Operated by the National Railroad Passenger Corporation of Washington, DC. This rail system was created by President Nixon in 1970, and was given the responsibility for the operation of intercity, as distinct from suburban, passenger trains between points designated by the Secretary of Transportation.	**美国铁路客运公司**：由华盛顿特区全国铁路客运公司经营。这套铁路系统是由美国总统尼克松于 1970 年批准建立的；为不同城市间（而非城市市区和郊区之间）和交通部秘书指定的点对点的客运提供交通运输服务。
Anchor It: Apply brakes for an emergency stop.	**紧急刹车**：出现紧急情况时采取的制动措施.
Angle Collision: Collisions which are not head on, rear end, rear to rear, or sideswipe.	**角度碰撞**：那些除正面碰撞、追尾、尾尾碰撞和侧面刮擦之外的事故.
Annual Operating Factor: The annual fuel consumption divided by the product of design firing rate and hours of operation per year.	**年度运营指数**：每年的燃料消耗量除以设计耗油速率与年均运营小时数的乘机.
Antenna: A metallic apparatus for sending and receiving electromagnetic waves.	**天线**：发送和接收电磁波的金属装置.
Antenna Array: A group of directional antennas.	**天线组**: 一组定向天线.
Arch: A curved structure that supports the weight of material over an open space.	**穹顶**：用于承受开放空间上部顶棚的材料重量的弧形构造物。

English	Chinese
Arrangement of Passenger Transportation: Includes establishments engaged in providing travel information and acting as agents in arranging tours, transportation, car rentals, and lodging for travelers.	**客运安排**：包括提供出行资讯，代理旅游相关事宜，为旅行者提供交通，租车及住宿。
Arrangement of Passenger Transportation Not Elsewhere Classified: Establishments primarily engaged in arranging passenger transportation (other than travel agencies and tour operators), such as ticket offices (not operated by transportation companies) for railroads, buses, ships, and airlines.	**其他未归类的客运安排**：那些除旅行社和旅游经营者之外的提供客运服务的组织,如出售铁路、巴士、轮船和航空票的售票处(非交通运输公司经营)
Arterial Highway: (See also Freeway, Minor Arterial, Principal Arterial) Arterial highways serve major traffic movements or major traffic corridors. While they may provide access to abutting land, their primary function is to serve traffic moving through the area.	**干线公路**：(参见高速公路、次干路、主干路) 干线公路是主要的交通干道，为大规模快速的交通运输服务。虽然它们具有一定的可达性（即用路者可以到达临近用地），但其主要职能仍是便于交通穿越某一区域.
Arterial Street: A major thoroughfare, used primarily for through traffic rather than for access to adjacent land, that is characterized by high vehicular capacity and continuity of movement.	**主干路**：主要为过境交通服务，而非到达毗邻地区的主要通道,特点是较大的通行能力和连续的交通流。
Asphalt: A dark-brown-to-black cement-like material containing bitumens as the predominant constituents obtained by petroleum processing. The definition includes crude asphalt as well as the	**沥青**：暗棕色至黑色水泥状物质。主要含有从石油提炼出来的沥青。此定义包括粗沥以及以下成品: 水泥, 流质，乳化沥青乳液 (不含水), 及石油蒸馏沉

English	Chinese

following finished products: cements, fluxes, the asphalt content of emulsions (exclusive of water), and petroleum distillates blended with asphalt to make cutback asphalts.

淀和沥青质的混合物.

Assigned Vehicle: A vehicle provided to an organizational element of a government agency or contractor by General Services Administration's (GSA) Interagency Fleet Management System for a period of more than 30 days.

指定车辆：由总务管理处（GSA）下属的商用车管理系统分配给某政府机构或经营者的车辆，且租赁期至少 30 天

Associated Equipment: Any system, part or component of a boat as originally manu-factured or any similar part or component manufactured or sold for replacement, repair, or improvement of such system, part, or component; any accessory or equipment for, or appurtenance to, a boat; and any marine safety article, accessory, or equipment intended for use by a person on board a boat; but excluding radio equipment, as designated by the Secretary [of Transportation] under 46 U.S.C. 2101.

相关的设备：原装制造的船只某系统，零部件或任何用来替换、维修或改善这些部件而生产和销售的设备部件；任何船只上的配件或设备；任何供船上人员使用的航海安全用品，配件或设备；但是不包括由交通部文件 46U.S.C. 2101 中规定的无线电设备。

At Grade: See also Grade Crossings, Highway-Rail Crossing.

平面交叉：又见道路平面交叉口，公路铁路交叉口。

Automatic Fare Collection (AFC) System: The controls and equipment that automatically admit passengers on insertion of the correct fare in an acceptable form, which may be coins, tokens, tickets, or farecards (stored value farecards must be inserted again on exit, at

自动售检票 (AFC) 系统: 那些当乘客插入正确的票款时自动放行的检票窗口或设备。票款的形式包括：硬币,纪念品,入场券,或收费储值卡，储值卡必须在出口再次插入，并可能收取额外费用).

English	Chinese
which point an additional fare may be required).	
Automatic Restraint System: Any restraint system that requires no action on the part of the driver or passengers to be effective in providing occupant crash protection (e.g., air bags or passive belts).	**自动保护系统**：那些在碰撞中不需要乘客或驾驶员作出任何行动，而能为他们提供有效保护的系统 (例如, 气囊或被动式安全带).
Automatic Vehicle Location System: A system that senses, at intervals, the location of vehicles carrying special electronic equipment that communicates a signal back to a central control facility.	**车辆自动定位系统**：以一定时间间隔感应那些装有特殊电子装备车辆的位置，并向中心控制器传输信号的系统
Automatic Vehicle Monitoring System: A system in which electronic equipment on a vehicle sends signals back to a central control facility, locating the vehicle and providing other information about its operations or about its mechanical condition.	**车辆自动监控系统**：用于监控装有特殊电子设备车辆的系统。 装于此类车辆的特殊电子设备向硬件系统的中心控制器传送信号，可用于确定车辆位置及其他操作情况或车辆的机械状态。
Automobile: See also Bus, Car, Minivan, Motor Vehicle, Taxi, Vehicle.	**汽车**: 又见巴士,轿车,吉普车,机动车辆,出租车,汽车。
Automobile: Any 4-wheeled vehicle propelled by fuel which is manufactured primarily for use on public streets, roads, and highways (except any vehicle operated exclusively on a rail or rails), and that either 1) Is rated at 6,000 pounds gross vehicle weight or less; or 2) Which a) is rated more than 6,000 pound gross vehicle weight, but less than 10,000 pounds gross vehicle weight; b) is a type of vehicle for which the	**汽车**：四轮车，由燃料推动，主要用于公共街道、道路 与公路 (除任何仅在铁路钢轨上运行的车辆). 它可以是 1) 6000磅车辆总重量或以下;或 2) a) 车辆总重量大于六千磅但少于一万磅车辆总重量;b)那些全国公路交通安全管理局(NHTSA)依据 CFR 第 49 卷 523 规定的平均燃油经济性标准可以施行

English	Chinese
[National Highway Traffic Safety Administration (NHTSA)] Administrator determines, under paragraph b) of 49 CFR 523, average fuel economy standards are feasible; and c) is a type of vehicle for which the Administrator determines, under paragraph b) of 49 CFR 523, average fuel economy standards will result in significant energy conservation, or is a type of vehicle which the Administrator determines, under paragraph b) of 49 CFR 523, is substantially used for the same purposes as vehicles described in 1) above.	的车辆类型;和 c)根据 CFR 第 49 卷 523 条，平均燃油经济性标准将显著降低能源消耗或者是那些用途大致和 1)中描述车型相同的车型。
Automobile: A privately owned and/or operated licensed motorized vehicle including cars, jeeps and station wagons. Leased and rented cars are included if they are privately operated and not used for picking up passengers in return for fare.	**汽车**：私有和/或需颁发驾照才能驾驶的机动车辆，包括小轿车、吉普车和旅行车。对于那些租赁的车辆：如果是供私人使用，而非用来接载乘客以赚取 车费，也涵盖在'汽车'的定义之内。
Automobile Size Classification: Automobile size classifications as established by the Environmental Protection Agency (EPA)。	**小轿车大小分类**：小轿车大小分类是由环境保护局(EPA)设定的。
Automobile Transporter Body: Truck body designed for the transportation of other vehicles.	**运输汽车的车体**：用来运输其他车辆的卡车的车身部分。
Automotive Billing Module (AutoBill): This module creates non-GSA customer billing tapes and General Services Administration (GSA) interfund transactions from billing	**汽车计费模块 (AutoBill)**：这个模块产生非 GSA 客户帐单录音带和总务管理(GSA) interfund 交易帐单，记录产生

English	Chinese
records generated in the Transportation Interface and Reporting System (TIRES) and generates monthly accounting transaction information to send to the NEAR (National Electronic Accounting and Reporting) system.	的 运输界面和报告系统带来的每月会计交易信息，并将之送到附近(全国会计电算化 报道)系统。
Automotive Payment Module (AutoPay): This module processes all maintenance and extended warranty vendor invoices entered into the Fleet Service Station (FSS) Fleet Management System by the Maintenance Control Centers and processes the rental authorization records for commercial rent-a-car rentals from the Fleet Management Center.	**汽车付款模块 (AutoPay)：** 该模块处理车队管理系统里车队服务站(FSS)所有的维修及延长保修卖主发票。处理商业租车的出租授权。
Available Seat Mile: One seat transported one mile.	**可获得的座位里程:** 一个席位运送一英里
Available Ton Mile: One ton of capacity (passengers and cargo) transported one mile.	**可获得的吨英里 :** 一吨的运输量(乘客和货物)位移一英里.
Available Ton Miles: The aircraft miles flown in each inter-airport hop multiplied by the capacity available (in tons) for that hop.	**可获得的吨英里 :** 两个机场间的飞行距离（英里）乘以飞行该距离时装载的乘客和货物总重（吨）
Average Road Width: The average width of the travelway.	**平均道路宽度 :** 道路平均宽度。
Average Vehicle Fuel Consumption: A ratio estimate defined as total gallons of fuel consumed by all vehicles, divided by: 1) The total number of vehicles (for average fuel	**平均机动车燃油消耗 :** 一个估计出来的比值，具体定义为所有车辆消耗的燃料总加仑数除以：1) 车辆总数 (平均每车耗

English	Chinese
consumption per vehicle) or 2) The total number of households (for average fuel consumption per household).	油量)或 2)总户数 (平均每户耗油量)。

Average Vehicle Miles Traveled: A ratio estimate defined as total miles traveled by all vehicles, divided by: 1) the total number of vehicles (for average miles traveled per vehicle); or 2) the total number of households (for average miles traveled per household).

平均汽车行程：一个估计出来的比值，定义为所有车辆行驶的总英里数，除以：1)车辆总数(平均每车行程英里数)；或2)总户数(平均每户行程英里数)。

English	Chinese
Balanced Transportation: See Intermodalism (3).	平衡运输,见复合联运
Ballast: Material place on a track bed to hold the track in line and elevation and to distribute its load. Suitable material consists of hard particles (e.g., crushed rock, slag, gravel) that are stable, easily tamped, permeable, and resistant to plant growth.	道碴：放在道床上,用于固定轨道线形和高度,分散负荷的材料.该材料由稳定、易夯实、有渗透性、抗植物生长的硬质颗粒物组成(如碎石、矿渣和砾石)
Balloon Freight: Lightweight freight.	轻载货：重量轻的货物
Bareback: Tractor without its semitrailer.	单拖车：无半挂车的拖车
Base Period (Off-Peak Period): In transit, the time of day during which vehicle requirements and schedules are not influenced by peak-period passenger volume demands (e.g., between morning and afternoon peak periods). At this time, transit riding is fairly constant and usually low to moderate in volume when compared with peak-period travel.	基本周期(非高峰时段)：车辆在运行过程中,车辆需求和时间安排不受高峰期的客运量需求影响的时段(如在上午和下午的高峰期之间的时段).此时,客运量比较平稳,而且低于高峰期流量。
Basic Grant: The funds available to a State for carrying out an approved State Enforcement Plan (SEP), which include, but are not limited to: 1) Recruiting and training of personnel, payment of salaries and fringe benefits, the acquisition and maintenance of equipment except those at fixed weigh scales for the purposes of weight enforcement, and reasonable overhead costs needed to operate the program; 2) Commencement and conduct of expanded systems of enforcement; 3) Establishment of	基本基金: 一个州可用于实施已通过的州立执行计划(SEP)的资金。其中包括(但不限于)：(1)招募和培训人员,支付工资和福利,购置和维修设备(检查载重所用的间接费用)；(2)启动和实施扩大的执行体系；(3)建立有效的服务和遵守执行的制度,以及(4)人员和设备再培训与更新。

English	Chinese
an effective out-of-service and compliance enforcement system, and 4) Retraining and replacing staff and equipment.	
Beginning Milepost: The continuous milepost notation to the nearest 0.01 mile that marks the beginning of any road or trail segment.	**起始里程碑**: 指示公路或步道的起点的连续里程标志，其精度为 0.01 英里。
Bible: The "Golden Rule" safe driving book.	**圣经**:"黄金法则"安全驾驶手册。
Bicycle: A vehicle having two tandem wheels, propelled solely by human power, upon which any person or persons may ride.	**自行车, 脚踏车**, 一种具有两个串连轮子的车,单纯由人力驱动，可由一人或数人骑驶
Bicycle Lane (Bike Lane): A portion of a roadway which has been designated by striping, signing and pavement markings for the preferential or exclusive use of bicyclists.	**自行车道**：公路上由标志线,道路标志和路面标记等划分出的，供自行车优先行驶或专用的道路部分。
Bicycle Path (Bike Path): A bikeway physically separated from motorized vehicular traffic by an open space or barrier and either within the highway right-of-way or within an independent right-of-way.	**自行车专用道**: 由开放空间或障碍物与机动车道路分隔的自行车专用道。 可能与公路共享路权，或者具有独立的路权。
Bicycle Route (Bike Route): A segment of a system of bikeways designated by the jurisdiction having authority with appropriate directional and information markers, with or without a specific bicycle route number.	**自行车路网**: 权威管辖机构所指定的自行车道路系统，具有适当的指示和信息标志, 有或没有特定的自行车路线编号。
Bicycles: Includes bicycles of all speeds and sizes that do not have a motor.	**自行车**：非机动车辆, 没有马达的各种速度和大小的自行车。

English	Chinese
Big Hat: State Trooper.	大帽子：国家警察
Big Rigger: Arrogant driver, or one who will drive only long trailers.	牛司机：过度自信的司机,或长拖车司机
Bikeway: Any road, path, or way which in some manner is specifically designated as being open to bicycle travel, regardless of whether such facilities are designated for the exclusive use of bicycles or are to be shared with other transportation modes.	自行车道: 以某种方式为自行车行驶而指定的道路, 无论该道路是专供自行车行驶还是与其它交通工具共享。
Binders: Brakes.	闸: 刹车
Birdyback: Intermodal transportation system using highway freight containers carried by aircraft.	禽背：公路与航空复合运输,飞机托运载货挂车,由公路卡车驶进机舱,飞机卸货时(直接驶离,并将货物 运输送达目的地。
Blanket Certificate (Authority): Permission granted by the Federal Energy Regulatory Commission (FERC) for a certificate holder to engage in an activity (such as transportation service or sales) on a self implementing or prior notice basis, as appropriate, without case-by-case approval from FERC.	总证书 (权力): 由联邦能源管理委员会(FERC) 批准,为证书持有人在自执行或事先通知的基础上从事任何活动(如 运输或销售等), 适当的情况下, 联邦能源管理委员会不逐案审批。
Blood Alcohol Concentration (BAC): Is measured as a percentage by weight of alcohol in the blood (grams/deciliter). A positive BAC level (0.01 g/dl and higher) indicates that alcohol was consumed by the person tested. A BAC level of 0.10 g/dl or more indicates that the person was intoxicated.	血液酒精浓度: 血液中酒精百分比 (克/十分之一公升). 使酒精测试呈阳性的血液酒精浓度为 0.01 克/0.1 公升或以上 , 表示被测人摄入酒精。血液酒精浓度为 0.10 克/1/10 公升或以上则表示被测人已醉酒.

English	Chinese
Bodily Injury: Injury to the body, sickness, or disease including death resulting from any of these.	**人身伤害**: 身体的受伤,生病,或疾病，以及这些原因导致的死亡
Body: See also Chassis.	**车身**: 见底盘
Body: Semitrailer.	**车身**: 半挂车
Body Type: Refers to the individual classifications of motor vehicles by their design structure based on definitions developed by the Society of Automotive Engineers.	**汽车车型**: 指的是基于汽车工程师协会定义的以设计结构划分的汽车分类
Body Type: 1) The appearance of the vehicle. 2) Detailed type of motor vehicle within a vehicle type.	**汽车车型**： 汽车的外观. 2) 各类汽车的具体车型。
Bogey: An assembly of two or more axles.	**博忌**：两个或两个以上车轴装配
Boll Weevil: A novice truck driver.	**新手**：新卡车司机。
Boom It Down: Tighten chains around freight.	**绑定**：收紧货物链。
Boomers: Binder devices used to tighten chains around cargo on flatbed trailers.	**紧绳器**: 用于收紧平板拖车货物链的收紧装置。
Border Cargo Selectivity (BCS): An automated cargo selectivity system based on historical and other information. The system is designed to facilitate cargo processing and to improve Customs enforcement capabilities by providing targeting information to border locations. The system is used for the land-border environment.	**边境货物选择性（BCS）**：基于历史及其它资料的自动货运选择系统。 该系统旨在通过向边境提供信息以促进货运过程，并提高海关执法能力.该系统用于陆地边界。

English	Chinese
Bottlers Body: Truck body designed for hauling cased, bottled beverages.	**灌注机装载车**：专门运输瓶装，罐装饮料的车型
Bottom Dumps: Trailer that unloads through bottom gates.	**底卸拖运车**: 通过底门卸货的挂车
Boundary: A nonphysical line indicating the limit or extent of an area or territory.	**边界**: 指定一个地区或领土的限制范围的非自然线
Box: 1) Semitrailer; 2) The transmission part of the tractor.	**箱体**：1)半挂车, 2)挂车变速器:
Break the Unit: Uncouple the tractor from the trailer.	**拆散单位**: 将挂车从拖车上卸下。
Bridge: A structure including supports erected over a depression or an obstruction, such as water, highway, or railway, and having a track or passageway for carrying traffic or other moving loads, and having an opening measured along the center of the roadway of more than 20 feet between undercopings of abutments or spring lines of arches, or extreme ends of openings for multiple boxes; it may also include multiple pipes, where the clear distance between openings is less than half of the smaller contiguous opening.	**桥梁**: 建立在下沉地区或障碍物，如，水，高速公路，铁路，有轨交通或走廊上方的结构物。 其目的是用于车辆行走或是运输物资，沿道路中心线，从临界线或拱形的弹簧线，或多个盒子的边缘有超过20英尺。此概念也可以由多个管道构成，其标准为开放处的距离小于最小的连接处距离的一半。
Bridge Foundation Bearing Material: The type of material supporting the substructure of a bridge. Code as follows: GW-well graded gravel, GP-poorly graded gravel, GM-silty gravel, GC-clay gravel, SW-well graded sand, SP-poorly graded sand, SM-silty sand, SC-clay sand, RK-bedrock,	**桥基轴承材料**：支持桥梁结构的材料。编码如下：GW-分级良好的碎石。GP-分级不好的碎石。GM-充满淤泥的碎石。GC-粘土碎石，SW-分级良好的沙，SP-分级不好的沙，SM-充满淤

English	Chinese
UK-unknown, O-other.	泥的沙，SC-粘土沙，RK-岩床，UK-未知，O-其他。

Bridge Number: The number of the installation, consisting of the full route number (including segment and spur) plus the milepost location of the bridge to the nearest one hundredth of a mile.

桥梁编号：安装号码，包括完整的道路号码（包括路段，分支）加上桥梁里程碑到最近的白分之一英里。

Bridge Posted Load Restrictions: Load restrictions posted at a bridge structure. Entry order: single axle, dual axle, load type 3, load type 3S2, load type 3-3 and Special.

桥梁载重限度示：桥梁结构上标明的负载限制。顺序：单轴，双轴，负载类型3，负载类型3S2，负载类型3-3和特殊类型。

Bridge Posted Speed Restrictions: A speed limit posted at a bridge structure, in miles per hour.

桥梁限速标示：桥梁结构上的限速标志，每小时英里数。

Bridge Structure: A two character code for recording the type of bridge structure. Code as follows: SS-simple span, CS-continuous span, SC-combination simple and cantilever, CC-combination continuous and cantilever, O-other.

桥梁结构：两位数的代码，用于记录桥梁结构的类型。有如下代码：SS-单跨，CS-连续跨度，SC-组合单悬桁，CC-组合连续悬桁，O-其他。

Bridge Superstructure: Those elements of the bridge structure which are above the uppermost deck.

桥梁上部结构：在桥梁最上部甲板上方的桥梁结构。

Brownie: Auxiliary transmission.

布朗尼: 辅助传动系统

Bulk Cargo: Cargo not packaged or broken into smaller units. Bulk cargo is either dry (grain) or liquid (petroleum) and cannot be counted.

大宗货物: 没有包装的或是散落成小件的货物. 可以是干的(如谷物), 也可以是液体(如原油),不可数的货物.

English	Chinese

Bulk Terminal: A facility used primarily for the storage and/or marketing of petroleum products, which has a total bulk storage capacity of 50,000 barrels or more and/or receives petroleum products by tanker, barge, or pipeline.

散货码头: 主要用于储存和销售石油产品的设施。总储藏量至少为 5 万桶。通过油轮，驳船或管道来运输石油。

Bull Hauler: One who hauls livestock.

运牛者: 运输家畜的人.

Bumble Bee: A two-cycle engine.

弄糟蜂: 二冲程发动机

Bumpers: (See also Possum) 1) Fenders; 2) Pads made out of Styrofoam, old ropes, old tires, or similar material, which are hung over the side of a water vessel to prevent damage to the vessel when berthing or locking through dams.

保险杠, 减震器 , 缓冲器：1）挡泥板；2）泡沫聚苯乙烯，旧绳索，旧轮胎或相似的材料造的垫片，挂在船外边以防轮船停泊或通过大坝时轮船损伤。

Bunker: A storage tank.

沙坑： 储藏库.

Bureau of Transportation Statistics (BTS): The Bureau was organized pursuant to section 6006 of the Intermodal Surface Transportation Efficiency Act (ISTEA) of 1991 (49 U.S.C. 111), and was formally established by the Secretary of Transportation on December 16, 19BTS has an intermodal transportation focus whose missions are to compile, analyze and make accessible information on the Nation's transportation systems; to collect information on intermodal transportation and other areas; and to enhance the quality and effectiveness of DOT's statistical programs through research, the development of guidelines, and the promotion of

交通统计局：根据 1991 年复合地面联运效率法案 6000 条成立的机构。 在 1992 年 12 月 16 号由交通部正式组建。运输统计局侧重于多种运输方式，其任务是编辑，分析，和公布国家运输系统的信息。收集多种运输方式和其他的信息， 通过研究加强交通部的统计数量和有效性。开发指导方针，改进数据询问和使用。统计局由复合地面联运效率法案管理，分为 6 个功能领域， 1）编辑，分析，公布统计数据；2）开发长期数据收集程序；3）制定指南以改善统计

English	Chinese

improvements in data acquisition and use. The programs of BTS are organized in six functional areas and are mandated by ISTEA to: 1) Compile, analyze, and publish statistics 2) Develop a long-term data collection program 3) Develop guidelines to improve the credibility and effectiveness of the Department's statistics 4) Represent transportation interests in the statistical community 5) Make statistics accessible and understandable and 6) Identify data needs.

工作的可信度和有效性；4）代表统计届对交通的兴趣；5）使统计信息可获取和理解；6）指明数据需要。

Bus: Includes intercity buses, mass transit systems, and shuttle buses that are available to the general public. Also includes Dial-A-Bus and Senior Citizen buses that are available to the public.

公共汽车：包括城际，城区，班车，所有对公众开放的公共汽车。还包括传呼公共汽车，以及老年人公交车.

Bus Charter Service (Except Local): Establishments primarily engaged in furnishing passenger transportation charter service where such operations are principally outside a single municipality, outside one group of contiguous municipalities, or outside a single municipality and its suburban areas.

巴士包车服务 (不包括短途服务)：给旅客提供客运服务的机构，这些服务通常在一个市/镇以外，不局限单一的地区及周边地区。

Bus Lane: A street or highway lane intended primarily for buses, either all day or during specified periods, but sometimes also used by carpools meeting requirements set out in traffic laws.

公共汽车专用车道: 街道或高速公路上专门给公共汽车使用的车道.可以全天也可以在某些特殊时段. 有些时候也可以根据交通法规的要求给多人搭乘的小轿车使用.

Business District: The territory contiguous

商业区：临近并包括高速公路

English

to and including a highway when within any 600 feet along such highway there are buildings in use for business or industrial purposes, including but not limited to hotels, banks, or office buildings which occupy at least 300 feet of frontage on one side or 300 feet collectively on both sides of the highway.

Chinese

的地区, 尤其是位于公路六百英尺范围内商业或工业用途的建筑, 包括但不局限于酒店, 银行, 办公大楼. 这些建筑至少占据公路一边或两边三百英尺.

English	Chinese
Cab: Portion of truck where the driver sits; tractor. The passenger compartment of a vehicle.	**驾驶室**: 卡车上司机所在的部分。车辆的乘客车厢
Cab Over: A vehicle with a substantial part of its engine located under the cab. Also known as snubnose.	**平头车**: 一种车辆，其相当部分的引擎设置在驾驶室下方. 又名仰鼻。
Cab-Over-Engine (COE): A truck or truck-tractor, having all, or the front portion, of the engine under the cab.	**平头式（COE）**：一种卡车或牵引车，其引擎的全部或前部设置在驾驶室下方。
Cab-Over-Engine (COE) High Profile: A COE having the door sill step above the height of the front tires.	**平头式高侧面**：门槛踏板高于前排轮胎的平头式车。
Cab Signal: A signal located in engineman's compartment or cab, indicating a condition affecting the movement of a train and used in conjunction with interlocking signals and in conjunction with or in lieu of block signals.	**机车信号**: 位于火车司机隔舱或驾驶室内的一种信号，显示出影响火车运动的某种状况；与互锁信号联合使用，有时可代替地面信号。
Cab-To-Axle Dimension (CA): The distance from the back of a truck cab to the center line of the rear axle. For trucks with tandem rear axles, the CA dimension is given midway between the two rear axles.	**驾驶室-轴距离**：从卡车驾驶室的后面至后轴的中心线的距离. 对于有串联后轴的卡车, 该尺寸使用的后轴中心线取两后轴之间的中点。
Cackle Crate: Truck that hauls live poultry.	**家禽车**: 运输活家禽的卡车。
Camel Back Body: Truck body with floor curving downward at the rear.	**骆背卡车**:车体地板后部向下弯曲的卡车。
Camp Car: Any on-track vehicle, including outfit, camp, or bunk cars or modular homes mounted on flat cars used to house rail employees. It does not include wreck trains.	**住宿车厢**：任何带有营地设备，帐篷，铺位车厢或者移动住宅、安放在平台车上、专供修路或养路工人住宿的火车车厢。救援列车不包括在内。

English	Chinese

Cancellation of Insurance: The withdrawal of insurance coverage by either the insurer or the insured.

保险撤消：由投保人或者被保人作出的撤出保险的行为。

Car: See also Automobile, Minivan, Motor Vehicle, Taxi, Vehicle.

汽车: 参见汽车, 面包车, 机动车, 出租车, 车辆。

Car: 1) Any unit of on-track equipment designed to be hauled by locomotives; 2) Any unit of on-track work equipment such as a track motorcar, highway-rail vehicle, push car, crane, ballast tamping machine, etc; 3) A railway car designed to carry freight, railroad personnel, or passengers. This includes boxcars, covered hopper cars, flatcars, refrigerator cars, gondola cars, hopper cars, tank cars, cabooses, stock cars, ventilation cars, and special cars. It also includes on-track maintenance equipment.

车厢: 1) 任何由机车牵引的，运行于轨道上的各种车辆设备; 2) 任何于轨道上工作的设备如轨道车, 公路铁路两用车, 手推车，起重机，和捣碴机等; 3) 用以装载货物,铁路人员或乘客的铁路车辆,包括有盖货车、有盖漏斗车、平台型货车、冷藏车、敞车、漏斗车、罐车、守车、运牲口车、通风车和专用车，也包括轨载养路设备

Carburetor: (See also Diesel Fuel System, Fuel Injection) A fuel delivery device for producing a proper mixture of gasoline vapor and air, and delivering it to the intake manifold of an internal combustion engine. Gasoline is gravity fed from a reservoir bowl into a throttle bore, where it is allowed to evaporate into the stream of air being inducted by the engine. The fuel efficiency of carburetors is more temperature dependent than fuel injection systems.

化油器：（参见柴油机燃油系统，喷油）
一种燃料输送装置，用于产生适当的汽油蒸气和空气混合物，并将其运送到内燃机的进气总管.汽油靠重力从储油碗流进节流阀孔，在那里蒸发进入被发动机吸入的空气流中。与燃油喷射系统相比，化油器的燃料效率更加依赖于温度。

Cargo Insurance and Freight (CIF): Refers to cargos for which the seller pays for the transportation and insurance up to the port of destination.

货物运输保险及运输(到岸价):
指为由卖方支付到达目的地港口的运输和保险费用的货物.

English	Chinese

Cargo Tank: A bulk packaging which: 1) Is a tank intended primarily for the carriage of liquids or gases and includes appurtenances, reinforcements, fittings, and closures; 2) Is permanently attached to or forms a part of a motor vehicle, or is not permanently attached to a motor vehicle but which, by reason of its size. Construction or attachment to a motor vehicle is loaded or unloaded without being removed from the motor vehicle; and 3) Is not fabricated under a specification for cylinders, portable tanks, tank cars, or multi-unit tank car tanks.

货舱: 一个大容积的组合体，它: 1) 是一个罐子，主要用于运输液体或气体，包括附属物, 增强部件, 配件 和封闭体等; 2) 永久性地附着于一辆机动车，或成为其组成部分，或者不是永久性地附着于机动车, 但因其尺寸在装货或卸货时不离开机动车；3) 不是根据储筒、便携罐、罐车、 或是多单元罐车用罐的规格来制造的。

Cargo Tank Motor Vehicle: A motor vehicle with one or more cargo tanks permanently attached to or forming an integral part of the motor vehicle.

货物罐汽车: 永久性地附带着一个或多个货罐的汽车，或者这些货罐已成为该车不可分割的一部分

Cargo Ton-Miles: One ton of cargo transported one mile.

货运吨英里: 一吨的货物运送一英哩。

Cargo Transfer System: A component, or system of components functioning as a unit, used exclusively for transferring hazardous fluids in bulk between a tank car, tank truck, or marine vessel and a storage tank.

货物转运系统: 专门用于在罐、罐货车 或海洋船只与储罐之间转移危险液体的组件或系统组件单元。

Cargo-Carrying Unit: Any portion of a commercial motor vehicle (CMV) combination (other than a truck tractor) used for the carrying of cargo, including a trailer, semi trailer, or the cargo-carrying section of a single-unit truck.

货物运载单元: 用于运载货物的商业机动车 (CMV) 组合(除了卡车拖拉机以外)的任何部分。包括拖车、半挂车、或者单体货车的货物运载单元。

Carload: Shipment of freight required to fill a rail car.

货车载重量: 填满一辆铁路货车的货物运送量.

English	Chinese

Carpool: An arrangement where two or more people share the use and cost of privately owned automobiles in traveling to and from pre-arranged destinations together.

共乘: 两个或更多的人共同使用同一私人汽车往返于预先安排好的目的地，并分摊成本。

Carrier: A person engaged in the transportation of passengers or property by land or water, as a common, contract, or private carrier; or civil aircraft.

承运者: 从事运送乘客或物品的水、陆运输者。承运人可以是公共、契约或私营的; 也指民用航空器。

Carrier Liability: A common carrier is liable for all loss, damage, and delay with the exception of act of God, act of a public enemy, act of a public authority, act of the shipper, and the inherent nature of the goods. Carrier liability is specified in the terms of the bill of lading.

承运者责任: 一般承运人需对货物的一切丢失, 损害和拖延承担责任。除非货损是由于不可抗力、敌对势力、公共权力机关 或托运人所为和货物的内在特性所引起。 承运人责任在提单条款上指明

Carrier Type: "For-hire", private or "other."

承运者类型:"租用",私人或"其他" .

Casualty: See also Accident, Collision, Crash, Derailment, Fatality, Event, Incident, Injury, Personal Casualties terms.

伤亡: 参见事故，碰撞,坠毁,出轨,事故死亡,事件，受伤,和人身伤亡等词条。

Cement Mixer: Truck with a noisy engine or transmission.

水泥搅拌车: 带噪音发动机或传输器的卡车.

Census: The complete enumeration of a population or groups at a point in time with respect to well-defined characteristics: for example, population, production, traffic on particular roads. In some connection the term is associated with the data collected rather than the extent of the collection so that the term sample census has a distinct meaning. The partial enumeration resulting from a failure to cover the whole population,

人口普查：在一个时间段上，对具备明确定义特点的人群或团体作完整的计数: 譬如：人口、产量和特定道路上的交通量。在某些关系上，该词与收集的数据相关，而不是与收集的程度相关，因而抽样人口普查一词具有独特的意义。与设计好的抽样调查不同，由于未能复盖整个人口而只获得部分

English	Chinese

as distinct from a designed sample enquiry, may be referred to as an "incomplete census."

计数结果，可称作"不完整的人口普查."

Census Division: A geographic area consisting of several States defined by the U.S. Department of Commerce, Bureau of the Census. The States are grouped into nine divisions and four regions.

人口普查分区：一个地理区域，美国商务部人口普查局将全国各州分为九个分区,四大区域。

Central Business District (CBD): The downtown retail trade and commercial area of a city or an area of very high land valuation, traffic flow, and concentration of retail business offices, theaters, hotels and services.

中央商务区(CBD): 城市中心的零售贸易和商业区，或是具有有极高土地价值，极高交通流量，并集中了零售商业楼，影剧院，宾馆和服务设施的区域。

Central City: (See also Metropolitan Statistical Area, Standard Metropolitan Statistical Area) Usually one or more legally incorporated cities within the Metropolitan Statistical Area (MSA) that is significantly large by itself or large relative to the largest city in the MSA. Additional criteria for being classified as "Central City" include having at least 75 jobs for each 100 employed residents and having at least 40 percent of the resident workers employed within the city limits. Every MSA has at least one central city, usually the largest city. Central cities are commonly regarded as relatively large communities with a denser population and a higher concentration of economic activities than the outlying or suburban areas of the MSA. "Outside Central City" are those parts of the MSA not designated as central city.

中心城市：(参见大都市统计区,标准都市统计区) 通常指大都会统计区内一个或多个合法注册的、规模较大的城市。可列为中心城市的其他标准包括：相对每 100 名就业居民，城市至少有 75 个职位，而且受雇居民中至少有 40%在城市范围内就业。每个大都市统计区（MSA）至少有一个中心城市，通常是其中最大的城市。中心城市通常被视为比较大的社区，并有比 MSA 外围或郊区更高的人口聚集和经济活动的密集度。"中心城市外"是指 MSA 中未被定为中心城市的那些部分。

English	Chinese
Certificate of Inspection: A document certifying that merchandise (such as perishable goods) was in good condition immediately prior to shipment. Preshipment inspection is a requirement for importation of goods into many developing countries.	**检验证书**: 证明商品(如易腐货物等) 在即将装运前状况良好的文件。许多发展中国家都要求进口货物必须作运前检查。
Certification of Public Road Mileage: An annual document (certification) that must be furnished by each state to Federal Highway Administration (FHWA) certifying the total public road mileage (kilometers) in the state as of December 31 of the preceding year.	**公路里程认证**：每年各州必须向联邦公路管理局（FHWA）提交的文件（认证文件），认证截至 12 月 31 日该州内总公路里程(公里)的文件或证书。
Charge It: Let brake air flow into semi trailer lines.	**充气**：让刹车气流进入半挂车管线中。
Charter Service: A commercial passenger vehicle trip not scheduled, but specially arranged. The charter contract normally commits the carrier to furnish the agreed to transportation service at a specified time between designated locations.	**包租服务**：非预定而特殊安排的商业客运车辆出行。包租合同通常限定承包人必须在指定时间里，在指定地点之间，提供商定的运输服务。
Charter Service Hours: The total hours traveled/operated by a revenue vehicle while in charter service. Charter service hours include hours traveled/operated while carrying passengers for hire, plus associated deadhead hours.	**包租服务时间**：包租服务中营运车辆的总的旅行/运行小时数。包租服务时间包括租用时的载客旅行/运行时间和相关的空驶时间。
Charter Transportation of Passengers: Transportation, using a bus, of a group of persons who pursuant to a common purpose, under a single contract, at a fixed charge for the vehicle, have acquired the exclusive use of the vehicle to travel together	**旅客包租运输**：使用公共汽车运送一群人。这些人有共同的目的地，参与同一合同，接受固定的用车费，获得了车辆的专用权，按旅程表共同旅行。旅程表可以事先定好，也可在

English	Chinese
under an itinerary either specified in advance or modified after having left the place of origin.	离开出发地后再作修改。
Chassis: See also Body.	**底盘**：参见车体。
Chassis: The load-supporting frame in a truck or trailer, exclusive of any appurtenances which might be added to accommodate cargo.	**底盘**：卡车或拖车上的载物支撑框架，不包括任何为运输货物而添加的附属装置。.
Conflicting Movement: Movements over conflicting routes.	**冲突运动**: 在有冲突的路线上移动. 。
Conflicting Routes: Two or more routes, opposing, converging or intersecting, over which movements cannot be made simultaneously without possibility of collision.	**冲突的路线:** 两条或两条以上的路线,反方向,趋同或交叉。那些不能同时移动而不发生碰撞的路线.
Consist: On-track railroad equipment such as a train, locomotive, group of railcars, or a single railcar not coupled to another car or to a locomotive.	**车辆组合**: 有轨铁路设备，如火车机车,火车头, 或一个单一的独立机车.
Consist Responsibility: The railroad employing the crew members operating the consist at time of the accident determines the consist owner for reporting purposes only.	**车组责任**: 铁路雇用人员操作车组。 在事故发生时，用于报告车主的责任。.
Consolidated Metropolitan Statistical Area (CMSA): A metropolitan complex of 1 million or more population, containing two or more component parts designated as primary metropolitan statistical areas (PMSAs).	**合并大都市统计区(cmsa)**:都市人口１百万以上的地区, 包括两个或两个以上的主要都市统计区 (pmsas).
Consolidated Vehicle: A vehicle transferred, with or without reimbursement,	**联合汽车**: 通过加入初始车队管理系统， 汽车由一个政府部

English	Chinese
to General Services Administration (GSA) by another government agency for participating in the Introductory Fleet Management System (IFMS).	门,将车辆有偿或无偿转让给其他政府部门的通用服务属（GSA）。
Constant Dollars: A dollar value adjusted for charges in the average price level. A constant dollar is derived by dividing a current dollar amount by a price index. The resulting constant dollar value is that which would exist if prices had remained at the same average level as in the base period.	**恒值美元**: 根据平均价格水准调节后的美元价值. 恒值美元的价格是由现值美元的价格除以价格指数所得。恒值美元价格代表着如果价格保持与基础时期一致的恒定货币价值.
Construction/Maintenance Zone: An area, usually marked by signs, barricades, or other devices indicating that highway construction or highway maintenance activities are ongoing.	**建筑/维修区**:在一个地区,通常由标志,路障或其他装置标识出来,,表明公路建设或公路养护活动正在进行.
Consumer Commodity: A material that is packaged and distributed in a form intended or suitable for sale through retail sales agencies or instrumentalities for consumption by individuals for purposes of personal care or household use. This term also includes drugs and medicines.	**消费商品**: 包装起来并以准备或适合通过零售商销售的形式配送, 以消费为目的的个人护理或家庭使用. 这个名词也包括药品.
Consumer Complaint: Oral or written communication from a consumer indicating a possible problem with a product.	**消费者投诉**: 消费者口头或书面的沟通,表示某个产品可能出现的问题。
Consumer Price Index (CPI): An index issued by the U.S. Department of Labor, Bureau of Labor Statistics. The CPI is designed to measure changes in the prices of goods and services bought by wage earners and clerical workers in urban areas.	**居民消费价格指数(CPI)**: 由美国劳工部劳工统计局公布的指数. 消费物价指数是用来衡量城市工薪阶级和办事员购买的商品和服务价格。它代表了典型消费的成本和其在 基期成本

English	Chinese
It represents the cost of a typical consumption bundle at current prices as a ratio to its cost at a base year.	的对照比率.
Consumer Product Safety Act: Establishes the Consumer Product Safety Commission. Definition of consumer product does not include boats which are covered under the statutes.	**消费产品安全法**: 规定消费者产品安全职责。消费产品的定义不包括法规所涵盖下的游艇.
Consumption Unit Value: Total price per specified unit, including all taxes, at the point of consumption.	**消费单位净值**: 每指定的单位的产品总价, 包括了消费时所有的税收。
Container: A component other than piping that contains a hazardous fluid.	**集装箱**: 除了管道以外的组成部分,装有危险的液体。
Container Chassis: A semi trailer of skeleton construction limited to a bottom frame, one or more axles, specially built and fitted with locking devices for the transport of cargo containers, so that when the chassis and container are assembled, the units serve the same function as an over the road trailer.	**集装箱**: 一半挂车骨架建设局限于个底框, 一个或一个以上车轴, 专门修建并配备锁装置用于运输货物集装箱, 因此, 当底盘及货柜组装上, 此单元和公路上的拖车服务功能一样.
Containership: A cargo vessel designed and constructed to transport, within specifically designed cells, portable tanks and freight containers which are lifted on and off with their contents intact.	**集装箱船**: 一艘货船的设计和建造用专门设计的单元来运输, 便携式罐体和集装箱可被升降而不用触及内部装的货物.
Continent: One of the large, unbroken masses of land into which the Earth's surface is divided.	**大陆**: 地球表面划分出来的一个大的 , 未分割的土地。
Contract Carrier: For Hire interstate operators [which] offer transportation services to certain shippers under contracts.	**合同承运人**: 根据合同雇用州际经营商提供运输服务给运货人。

English	Chinese

Contract Demand: The level of service in terms of the maximum daily and/or annual volumes of natural gas sold and/or moved by the pipeline company to the customer holding the contract. Failure of a pipeline company to provide service at the level of the contract demand specified in the contract can result in a liability for the pipeline company.

合同需求: 服务水平，它基于每天最高或年度天然气销售量或管道公司根据合同向客户运送的天然气量。如果管道公司不能按合同确定的服务水平的需求提供服务，会导致管道公司承担责任。.

Contracted Gas: Any gas for which Interstate Pipeline has a contract to purchase from any domestic or foreign source that cannot be identified to a specific field or group. This includes tailgate plant purchases, single meter point purchases, pipeline purchases, natural gas imports, SNG purchases, and LNG purchases.

协议气量: 州际管道公司，持有合同,向任何国内或,不能确定某一特定领域或集团的国外来源购买天然气. 这包括栏板厂购货单表点购买,购买管道,天然气进口 sng 购买,以及购买液化天然气.

Contractor Employee: A person employed by a contractor hired by a railroad to perform normal maintenance work to railroad rolling stock, track structure, bridges, buildings, etc.

承包员工人数:受雇于铁路雇用承包商的工人，对铁路机车车辆 轨道结构, 桥梁及建筑物等进行正常的维修工作

Control Area: (As it pertains to the Interstate Highway System) A metropolitan area, city or industrial center, a topographic feature such as a major mountain pass, a favorable location for a major river crossing, a road hub which would result in material traffic increments on the Interstate route, a place on the boundary between two States agreed to by the States concerned, or other similar point of significance.

控制区: (涉及到州际公路系统)的大都市区, 城市和工业中心,地形特点,例如主要山口,一个有利位置的一大过河,公路枢纽,这将导致州际公路上材料交通量增加，两个州交界处。 由有关的州商定, 或其他重要程度类似的决定。

English	Chinese
Control Cab Locomotive: A locomotive without propelling motors but with one or more control stands.	**控制机车的车头**: 无驱动马达的机车, 但有一个或更多的控制停机.
Control Circuit: An electrical circuit between a source of electric energy and a device which it operates.	**控制电路**: 电源和它控制的设备之间的电路。
Control Machine: An assemblage of manually operated devices for controlling the functions of traffic control system; it may include a track diagram with indication lights.	**数控机床**: 用于控制交通控制系统的手动控制装置的集合; 它可能包括带显示灯的轨迹图。
Control Operator: An employee assigned to operate the control machine of a traffic control system.	**控制操作**: 员工被安排操作交通控制系统的控制机器.
Control Station: The place where the control machine of a traffic control system is located.	**控制站**: 放置交通控制系统的地方.
Control System: A component, or system of components functioning as a unit, including control valves and sensing, warning, relief, shutdown, and other control devices, which is activated either manually or automatically to establish or maintain the performance of another component.	**控制系统**: 一个组成部分或系统运作组件作为一个单元, 包括控制阀和传感装置,传感器, 报警器,消除, 关机,以及其他控制装置用于已于手工或自动地建立或保持另一组成部分的的工作.
Controllable Emergency: An emergency where reasonable and prudent action can prevent harm to people or property.	**可控的紧急情况**: 紧急情况下合理和明智的行动可以防止伤害人身或财产.
Controlled Point: A location where signals and/or other functions of a traffic control system are controlled from the control machine.	**控制点**:所在位置信号和/或其他功能的交通控制系统控制的数控机床.

English	Chinese
Controlling Locomotive: A locomotive arranged as having the only controls over all electrical, mechanical and pneumatic functions for one or more locomotives, including controls transmitted by radio signals if so equipped. It does not include two or more locomotives coupled in multiple which can be moved from more than one set of locomotive controls.	**控制机车**: 机车被安排做为唯一的控制一个或一个以上的机车所有电器, 机械及气动功能. 若有装备的话, 还包括控制传送无线电信号,. 它不包括两个或两个以上的机车由多重 机车控制.
Conventional Cab: A cab design in which the engine is located ahead, or mostly ahead, of the cowl.	**常规驾驶室**: 驾驶室的设计使得发动机处于机壳的前端, 或大部分, 在最前端.
Conversion Factor: (See also British Thermal Unit) A number that translates units of one system into corresponding values of another system. Conversion factors can be used to translate physical units of measure for various fuels into British Thermal Unit (BTU) equivalents.	**换算因子**: (参见英国热单位), 一个可将数目单位从一个系统转变成另一个系统的相应的值. 换算因素可以用来把 各种燃料的物理的度量单位转换成英国热单位(btu)当量.
Converter Dolly: A motor vehicle consisting of a chassis equipped with one or more axles, a fifth wheel and/or equivalent mechanism, and drawbar, the attachment of which converts a semi trailer to a full trailer.	**可转换的挂车**: 汽车组成的底盘上装有一个或一个以上车轴, 第五轮和/或类似机制, 与拉杆, 可将半挂车转换成全挂车.
Convertible: A truck or trailer that can be used either as a flatbed or open-top by removing side panels.	**敞篷车**: 一辆卡车或拖车, 将侧面板去掉, 就能转变成一个平底或敞蓬车.
Copter: See Helicopter.	**直升机**: 见直升机。
Corporate Average Fuel Economy (CAFE) Standards: CAFE standards were originally established by Congress for new automobiles, and later for light trucks, in Title	**公司平均燃油经济(CAFE)标准**: CAFÉ 标准最初是由美国国会给新汽车设定, 后来是轻型卡车, 在车牌 v 车辆信息及成本

English	Chinese
V of the Motor Vehicle Information and Cost Savings Act (15 U.S.C. 1901, et seq.) with subsequent amendments. Under CAFE, automobile manufacturers are required by law to produce vehicle fleets with a composite sales-weighted fuel economy which cannot be lower than the CAFE standards in a given year, or for every vehicle which does not meet the standard, a fine of $5.00 is paid for every one-tenth of a mpg below the standard.	节约法(15u.s.c.1901 年, 7818)以后的修正案. 根据 CAFÉ 规定, 汽车厂商都须依法生产车队, 其燃油经济性, 不能比在某一年份的 CAFE 标准低. 如果车辆不符合标准, 低于标准每加仑公里数的十分之一的罚款五美元.
Cost, Insurance, Freight (CIF): A type of sale in which the buyer of the product agrees to pay a unit price that includes the f.o.b. value of the product at the point of origin plus all costs of insurance and transportation. This type of transaction differs from a delivered purchase in that the buyer accepts the quantity as determined at the loading port (as certified by the Bill of Lading and Quality Report) rather than pay on the basis of the quantity and quality ascertained at the unloading port. It is similar to the terms of an f.o.b. sale, except that the seller, as a service for which he is compensated, arranges for transportation and insurance.	**成本,保险费加运费价(到岸价格)**: 一类买卖中,买方同意支付的单价,包含产品在原产国的离岸 值, 加所有的保险和运输费用. 这类交易不同于购买运送到的商品, 买方接受 在装货港确定的数量(如提单和质量报告上标明),而不是根据卸货港卸货 数量和质量确定. 这是一个 f.o.b 的类似条款. 销售,除了卖方,安排运输和保险并获得相应补偿,.
Courier Services (Except By Air): Establishments primarily engaged in the delivery of individually addressed letters, parcels and packages (generally under 100 pounds), except by means of air transportation or by the U.S. Postal Service.	**速递服务(除空运)**: 机构,主要运送独立的信件,包裹(一般为一百磅以下),但用空运或由美国邮政服务的不包括在内.

English	Chinese

Coverage Ratios: The ratio used to measure the degree to which expenditures are funded or "covered" by the various types of revenues. This ratio indicates the percent of expenditures that is funded by identifiable transportation-related tax receipts, fees, etc.

复盖比率: 值用来衡量支出的资金被各类营业收入"覆盖"的程度. 这个比例是与运输有关的税务发票, 收费等等提供的支出比例.

Cowboy: Reckless driver.

牛仔: 鲁莽司机.

Crack: A fracture without complete separation into parts, except that castings with shrinkage cracks or hot tears that do not significantly diminish the strength of the member are not considered to be cracked.

裂纹: 裂缝没有完全分离成零件, 除了铸件收缩裂缝或不会大量降低未被撕裂的热撕拉。

Crash: An event that produces injury and/or damage, involves a motor vehicle in transport, and occurs on a traffic way or while the vehicle is still in motion after running off the traffic way.

碰撞: 造成人身伤害和/或损坏的事件, 涉及汽车运输, 发生在交通道路上 , 或在车辆偏离路面仍行驶的情况下。

Crash Severity: The most severe injury sustained in the crash as recorded on the police accident report: Property Damage Only (no injuries), Minor or Moderate (Evident, but not incapacitating; complaint of injury; or injured, severity unknown), Severe or Fatal (killed or incapacitating).

碰撞严重性: 在警察事故报告中最严重的受伤记录, 只是财产损坏 (无人受伤),轻微或中度 (显而易见的,但并非无所作为; 申诉人受伤; 或受伤,伤势不明), 重度或致命(死亡或致残).

Crash Type: Single vehicle or multiple vehicle crash.

碰撞类型: 单一车或多车碰撞.

Creeper Gear: Lowest gear or combination of gears used for extra power. Also known as Grandma.

爬行装置:最低齿轮或组合齿轮用于额外的动力. 又名老奶奶.

Crevasse: A deep fissure in snow or ice.

裂缝 : 雪或冰.表面的裂缝

English	Chinese

Crevasse Field: An area of deep fissures in the surface of an ice mass caused by breaking or parting.

决口领域: 冰表面的深刻裂痕, 由破坏或分离所造成.

Crewmember: A person assigned to perform duty in an aircraft during flight time.

队员: 飞机在飞行时, 在机舱内负责专工作的人.

Crib: A crate like construction of logs or beams, usually filled with stones, placed in water as a free standing mooring device or as the foundation of a pier or wharf.

木箱: 像箱子一样, 建造原木或横梁, 通常装满石头, 放置在水中作为一个自由常设泊装置或作为码头或码头的基础.

Crib Area: Containing one or more cribs, (frames of logs or beams filled with heavy material that are sunk and used as foundations or retaining walls for docks, piers or similar structures, or as supports for pipelines).

箱区: 含有一个或多个木箱, (框原木或横梁充满着沉重的物质可作为船坞,码头或类似机构, 或作为管道的支持基础或护墙).

Critical Altitude: The maximum altitude at which, in standard atmosphere, it is possible to maintain, a specified power or a specified manifold pressure at a specified rotational speed. Unless otherwise stated, the critical altitude is the maximum altitude at which it is possible to maintain, at the maximum continuous rotational speed, one of the following: 1) The maximum continuous power, in the case of engines for which this power rating is the same at sea level and at the rated altitude; 2) The maximum continuous rated manifold pressure, in the case of engines, the maximum continuous power of which is governed by a constant manifold pressure.

临界海拔: 标准大气可以维持的最高海拔, 即指定动力或在一个特定的转速下的多重压力的最高海拔高度。除非另有规定,临界海拔是最高连续转速可能维持的其中之一: 1) 当发动机的连续功率与海平面相同时, 连续输出的最大功率。 2) 当最大连续功率是由一个常数多重压力控制时, 发动机的最高连续多重压力。

Cross Protection: An arrangement to prevent the improper operation of a signal,

交叉保护: 防止不当操作信号开关, 可动心轨辙叉的措施.

English	Chinese

switch, movable-point frog, or derail as the result of a cross in electrical circuits.

或交叉电路导致的脱轨.

Cross-Servicing: Petroleum products, repairs, supplies, and services provided to General Services Administration (GSA) by other Federal agencies, or vice versa. Cross servicing may also refer to commercial firms where GSA or other Federal agencies have agreements with these firms to supply services repairs, or fuel.

交叉服务: 石油产品, 维修用品和服务提供给其他联邦机构的一般服务管理局, 反之亦然. 交叉服务也可指与联邦服务局或其他联邦机构有协议的商业公司, 公司给它们提供服务维修, 或燃油.

Crosstie: The transverse member of the track structure to which the rails are fastened. Its function is to provide proper gauge and to cushion, distribute, and transmit the stresses of traffic through the ballast to the roadbed.

枕木: 轨道结构的横着的部分. 铁轨固定在上面。. 它的功能是提供适当的规范与缓冲, 分配, 并把车辆的压力传递到路基上.

Crude Oil: A mixture of hydrocarbons that exists in the liquid phase in natural under-ground reservoirs and remains liquid at atmospheric pressure after passing through surface separating facilities.

原油: 混合烃类, 液态, 储于天然地下油库, 经过表面分离设施, 在大气压下还为液态.

Crude Oil Imports: The volume of crude oil imported into the 50 States and the District of Columbia, including imports from U.S. ter-ritories but excluding imports of crude oil into the Hawaiian Foreign Trade Zone.

原油进口: 原油量进口到 50 个州和哥伦比亚区, 包括来自美国的领土的进口, 但不包括进入夏威夷外国贸易区的进口原油.

Crude Oil Petroleum: A naturally occurring, oily, flammable liquid composed principally of hydrocarbons. Crude oil is occasionally found in springs or pools but usually is drilled from wells beneath the earth's surface.

原油石油: 一种自然发生的, 油性, 易燃液体, 主要包括碳氢化合物. 原油偶尔在喷泉里被发现, 但通常从地球表面钻探井下方发掘.

Crude Oil Production: The volume of crude oil produced from oil reservoirs during given

原油产量: 一段时间内原油产地的产量。 一段时间内的石油

English	Chinese

periods of time. The amount of such production for a given period is measured as volumes delivered from lease storage tanks (i.e., the point of custody transfer) to pipelines, trucks, or other media for transport to refineries or terminals with adjustments for 1) net differences between opening and closing lease inventories, and 2) basic sediment and water (BS&W).

产量由从租赁储油罐(即所有权移转) 运送管道,卡车 或其他媒体运送到炼油厂或终端的油量决定.可根据以下情况调整：) 进出合同清单的净差异,以及 2) 基本沉积物和水(bs&w).

Cruising: Proceeding normally, unrestricted, with an absence of drastic rudder or engine changes.

巡航: 照常进行,不受限制,没有剧烈 的方向舵或发动机的变化.

Cubic Inch Displacement (CID): A measure of the physical size of the engine.

立方英寸位移器(CID):量度发动机的实物大小.

Cul-De-Sac: The round or circular section of the end of a dead-end street.

巷尾: 死胡同的圆形或圆环形路段。

Curb Weight: The weight of a motor vehicle with standard equipment, maximum capacity of fuel, oil, and coolant: and, if so equipped, air conditioning and additional weight of optional engine. Curb weight does not include the driver.

路沿重量: 标准配备的汽车的重量,最大容量燃料,石油和冷却剂,若有配备空调和额外引擎的,它们的重量也应算上。curb 重量并不包括司机.

Current Assets: Cash and cash equivalents, as well as current receivables and short-term investments, deposits and inventories.

流动资产: 现金和现金等价物,以及当前应收款和短期投资,存款和库存.

Current Dollars: The dollar value of a good or service in terms of prices current at the time the good or service is sold. This contrasts with the value of the good or service measured in constant dollars.

当前价: 的美元价值的商品或服务的价格,目前在当时的商品或服务出售. 这与价值的商品或服务以不变价.

English	Chinese
Current Dollars: Represents dollars current at the time designated or at the time of the transaction. In most contexts, the same meaning would be conveyed by the use of the term "dollars".	**现价**: 代表美元当前在指定的时间,或在当时的交易. 在大多数情况下,同样的意思可传达所用的"元".
Current Liabilities: Current portion of long-term debt and of capital leases, air travel liabilities and other short-term trade accounts payable.	**流动负债**: 长期债务和资本契约,空中旅行负债及其他短期贸易帐款的当前部分.
Current of Traffic: The movement of trains on a specified track in a designated direction.	**交通流**: 火车在指定的轨道上,在一个指定的方向运动.
Cut: An excavation of the Earth's surface to provide passage for a road, railway, canal, etc.	**挖掘**: 挖掘地球表面,开发出马路,铁路,运河等做交通用途.
Cylinder: A pressure vessel designed for pressures higher than 40 psi and having a circular cross section. It does not include a portable tank, multi-unit car tank, cargo tank, or tank car.	**汽缸**:压力容器设计压力高于40psi,并具有圆截面. 它不包括便携的车 ,多组坦克车,货箱,或罐车.
Cylinder: (See also Engine Displacement, Engine Size) In a reciprocating engine, a cylinder is the chamber in which combustion of fuel occurs and the piston moves, ultimately delivering power to the wheels. Common engine configurations include 4, 6, and 8 cylinders. Generally, the more cylinders a vehicle has, the greater the amount of engine power it has. However, more cylinders often result in less fuel efficiency.	**汽缸**: (参见发动机位移,引擎马力)在交互机理 ,汽缸是燃烧运动和活塞动作发生的中庭 ,最终将动力传递到轮子上. 通用发动机配置,包括4 ,6 和8 缸. 一般而言,缸数越多 ,汽车产生的动力越大。不过 ,缸数的增加也会带来燃料效率的降低。

English	Chinese
Daily Average Flow: The volume of gas that moves through a section of pipe determined by dividing the total annual volume of gas that moves through a section of pipe by 365 days. Volumes are expressed in million cubic feet per day measured at a pressure of 14.73 psia and a temperature of 60 degrees Fahrenheit. For pipes that operate with bidirectional flow, the volume used in computing the average daily flow rate is the volume associated with the direction of flowing gas on the peak day.	**日均流量**: 通过一段管道的气体的量。一段管道气体的年总量除以 365 天。气体量的单位在测量压力 14.73 帕斯卡和温度 60 华氏度下是百万立方英尺单位。对于双向流通的管道，平均每日流量与高峰日期流动方向有关。
Daily Vehicle Travel: Is the amount of vehicle travel (in thousands) accumulated over a 24-hour day, midnight to midnight, traversed along a "public road" by motorized vehicles, excluding construction equipment or farm tractors. Vehicle travel not occurring on public roads, such as that occurring on private land roads (private roads in parking lots, shopping centers, etc.) must be also be excluded.	**车流量**：流量是指累计 24 小时，午夜十二时至午夜十二时，行驶在"公共道路"的汽车（千量计）。由机动车辆，但不包括建筑设备或农用拖拉机。汽车旅行未发生公共道路，例如像发生在私人土地上的道路（私人道路的停车场，购物中心等），也必须被排除在外。
Dead Axle: Non powered rear axle on tandem truck or tractor.	**静轴，从动轴**：串联的卡车或拖车上不带动力的后轴。
Dead Heading: Running empty.	**跑空车，放车**：跑空。
Defined Mandatory Use Service Area: That listed in the determination of each Fleet Management Center or Fleet Management Subcenter.	**规范使用服务区**：每个车队管理中心或车队管理下属中心确定的服务区。
Delivered Energy: The amount of energy delivered to the site (building); no adjustment	**输出的能源**：运送到工地（建设）的能源量，不包括消耗在

English	Chinese
is made for the fuels consumed to produce electricity or district sources. This is also referred to as net energy.	生产过程和产出地的燃料。也被称为净能源。

Department of Transportation (DOT): Establishes the nation's overall transportation policy. Under its umbrella there are ten administrations whose jurisdictions include highway planning, development and construction; urban mass transit; railroads; aviation; and the safety of waterways, ports, highways, and oil and gas pipelines. The Department of Transportation (DOT) was established by act of October 15, 1966, as amended (49 U.S.C. 102 and 102 note), "to assure the coordinated, effective administration of the transportation programs of the Federal Government" and to develop "national transportation policies and programs conducive to the provision of fast, safe, efficient, and convenient transportation at the lowest cost consistent therewith."

交通部：确立整个国家的整体交通政策的行政部门。在其组织下，有 10 个行政管理机构，包括公路规划、开发和建设、城市轨道交通、铁路、航空以及水路、港口、公路、石油和天然气管道的安全部门。交通部在 1966 年 10 月 15 日，依照法律（联邦法典第 102 和第 102 注）成立，其宗旨是"用于保证联邦政府协调，有效的管治交通项目"、"并制定"国家交通运输政策和方案，以最低的成本提供快捷、安全、高效和便捷的交通服务"。

Departure Angle: The smallest angle, in a plane side view of an automobile, formed by the level surface on which the automobile is standing and a line tangent to the rear tire static loaded radius arc and touching the underside of the automobile rearward of the rear tire.

错角：从汽车平面看，最小的角度。由汽车停立的水平面和线切线组成。该线正切于后轮轮胎静态负载半径圆弧，接触汽车的后方轮胎。

Derailment/Left Roadway: A non-collision incident in which a transit vehicle leaves the rails or road on which it travels. This also includes rollovers. Reports are made for all occurrences.

翻车，出轨：非撞击事件，车辆离开行驶的铁轨或道路，包括翻车。情况被记录为报告。

English	Chinese
Deregulation: Revisions or complete elimination of economic regulations controlling transportation. For example, the Motor Carrier Act of 1980 and the Staggers Act of 1980 revised the economic controls over motor carriers and railroads.	**放松管制**：修改或彻底消除经济法规对交通运输的控制。例如，1980 年机动车经营者法案和 1980 年的机动运输法案修订了国家对于机动运输工具和铁路的经济管制。
Design Capacity: (See also Certified Capacity) Is the capacity associated with the direction of the flow observed on the peak day.	**设计通行能力**：（亦见认证的能力），在繁忙时间（一天）、一个车流观察方向上的运输能力。
Design Speed: Design speed determines the maximum degree of road curvature and minimum safe stopping, meeting, passing, or intersection sight distance.	**设计时速**：设计速度决定了最大程度的道路曲率和最低安全停车、会车、超车、及交叉口视线距离。
Designated Area: The Fleet Management Center location as defined in the applicable determination.	**指定区域**：车队管理根据需要确定的中心选址。
Designated Facility: A hazardous waste treatment, storage, or disposal facility that has been designated on the manifest by the generator.	**指定设施**：由产生者在"自白书"中指定的危险废物处理、存储、使用或处置设施。
Designated Seating Capacity: The number of designated seating positions provided.	**指定的座位容量**：设施所能提供的指定座位的数量。
Designated Seating Position: Any plan view location capable of accommodating a person at least as large as a 5th percentile adult female, if the overall seat configuration and design and vehicle design is such that the position is likely to be used as a seating	**指定座位位置**：任何一个平面位置，大小至少可容纳一位成年女性。这样的整体座位配置与设计及车辆设计，能被用来作为座椅位置，汽车在运动时辅助座位可作为暂时或折叠跳

English	Chinese

position while the vehicle is in motion, except for auxiliary seating accommodations such as temporary or folding jump seats. Any bench or split-bench seat in a passenger car, truck or multipurpose passenger vehicle with a Gross Vehicle Weight Rating (GVWR) less than 10,000 pounds, having greater than 50 inches of hip room (measured in accordance with Society of Automotive Engineers (SAE) Standard J1100a) shall have not less than three designated seating positions, unless the seat design or vehicle design is such that the center position cannot be used for seating.

转席位。任何座席或分班座椅，在客车，卡车或多用途客运车辆与车辆总重评级少于一万磅，大于五十英寸的髋部室（实测按照国际汽车工程师学会（汽车工程学会）标准J1100a）须有不少于三个指定的座位位置，除非座椅设计或车辆设计是这样的：该中心的位置，不用作座位。

Designated Service: Exclusive operation of a locomotive under the following conditions; 1) The locomotive is not used as an independent unit or the controlling unit is a consist of locomotives except when moving for the purpose of servicing or repair within a single yard area; 2) The locomotive is not occupied by operating or deadhead crews outside a single yard area; and 3) The locomotive is stenciled "Designated Service-DO NOT OCCUPY".

指定服务：根据以下条件独家运营机车：1）该机车不是独立的单位或监管单位，作为机车时，在服务或修理之外，服务于单一的堆场；2）机车是否占用经营或船员以外的单一的堆场面积；3）机车是"指定的服务-不占用"。

Destination: For travel period trips, the destination is the farthest point of travel from the point of origin of a trip of 75 miles or more one-way. For travel day trips, the destination is the point at which there is a break in travel.

目的地，终点：对于一段时间的旅行来说，目的地是从出发地点到最远点，单程 75 英里或者更远。对于一日旅行，目的地是旅行的间歇点。

English	Chinese

Destination: The country in which the cargo was unloaded and/or the transit terminated.

目的地国：货物卸下和/或过境停留的国家。

Determination: A document signed by the Administrator of the General Services Administration, setting forth the decision to establish an Interagency Fleet Management Center at a specific location.

判定：由通用服务管理属主任签署，决定在某一特定地点设立一个车队管理中心的文件。

Direct Assistance: Transportation and other relief services provided by a motor carrier or its driver(s) incident to the immediate restoration of essential services (such as, electricity, medical care, sewer, water, telecommunications, and telecommunication transmissions) or essential supplies (such as, food and fuel). It does not include trans-portation related to long-term rehabilitation of damaged physical infrastructure or routine commercial deliveries after the initial threat to life and property has passed.

直接援助：由汽车承运人或其司机提供的交通运输和其他救济服务，以立即恢复基本服务（如，电力，医疗保健，污水处理，供水，电信，通信传输）或必需用品（如，食品和燃料）供给。它不包括最初威胁市民生命和财产的状况已经过去后，相关的长期恢复被破坏的物质基础设施或例行的商业运送。

Disabling Damage: Damage which precludes departure of a motor vehicle from the scene of the accident in its usual manner in daylight after simple repairs.

损伤：车辆承受损失的程度。即使经过简单行李也无法离开事故现场的损坏。

Disabling Injury: An injury causing death, permanent disability or any degree of temporary total disability beyond the day of the injury.

残障性伤害：造成死亡，永久性残废或在受伤日后任何程度的临时残疾的伤害。

Disc Brake: A brake used primarily on rail passenger cars that uses brake shoes clamped by calipers against flat steel discs.

盘式刹车：由弯脚器夹刹车片构成，主要用于铁路客运。

English	Chinese
Discontinued Operations Income (Loss): Gain or loss from disposal of investor controlled companies or non - transport ventures. Does not include earnings or losses from discontinued transport or transport-related operations.	**不再继续的收入(亏损)**：收益或损失由投资者控制的公司或非交通独资企业经营。不包括中止运输或与交通相关业务的盈利或亏损。
Dispatch Point: A location where arrangements may be made for the short term or trip rental use of an Interagency Fleet Management System (IFMS) vehicle.	**调度站，发送点**：安排运营业务的场所，可用于短期使用或出租的一个车队车辆管理系统。
Dispatch Vehicle: A vehicle provided to an individual or an organizational element of a government agency by GSA's Interagency Fleet Management System for short term use, usually on a day to day basis, not to exceed 30 consecutive days.	**调度车辆**：由联邦总务署的车队管理系统调制，提供给某一个人或一个组织的政府机构的车辆。为短期使用，在日常工作的基础上不超过三十天。
Disposal Date: The date a vehicle is disposed of and no longer included in the inventory.	**处置日期**：车辆停用并清除出库存的日期。
Disqualification: 1) The suspension, revocation, cancellation, or any other withdrawal by a State of a person's privileges to drive a commercial motor vehicle or 2) A determination by the Federal Highway Administration (FHWA), under the rules of practice for motor carrier safety contained in 49 CFR 386, that a person is no longer qualified to operate a commercial motor vehicle under 49 CFR 391; or 3) The loss of qualification which automatically follows conviction of an offense listed in 49 CFR 383.51.	**取消驾驶资格**：1）中止、撤销、注销，或其他任何方式取消一个人驾驶商业机动车的权利，或2）根据联邦公路管理局规则，汽车货物安全49篇386条，即一个人，不再有资格经营商业机动车，49篇391条；或3）根据49 383.51条款，失去驾驶资格。

English	Chinese

Distress: The state of being in peril, to any degree, for a person and/or property.

险况：人及/或财产正处于危难中，无论程度。

Distressed Unit: A person and/or property in peril to any degree.

遇险单位：一个人及/或财产处于危难中，无论程度。

Distribution Main: Generally, mains, services, and equipment that carry or control the supply of gas from the point of local supply to and including the sales meters.

配置总管：包括 维修、服务、设备，用于运送或控制天然气供应，从当地的供应到销售记量表。

Distributor: A company primarily engaged in the sale and delivery of natural and/or supplemental gas directly to consumers through a system of mains.

发行公司：公司主要从事销售和输送天然和/或补充气体，通过一个系统的主干，直接输送给消费者。

Ditch Light: Spotlight aimed at right side of road.

沟灯：道路右边的聚光灯。

Divided Highway: A multi-lane facility with a curbed or positive barrier median, or a median that is 4 feet (1.2 meters) or wider.

分隔式公路：一个多车道设施，带路沿或积极屏障隔离带，隔离带一般四英尺（1.2米），或更宽。

Division: An established point where driver ends trip.

分区： 司机完成旅程的结束点。

Dock It: Park truck at dock.

停车：把卡车停在码头。

Dock Receipt: A receipt used to transfer accountability when the export item is moved by the domestic carrier to the port of embarkation and left with the international carrier for export.

港站收据：一张收据，用来做会计转移，用于出口项目由国内运营商向港口上船，留给国际承运人以供出口。

Dock Walloper: One who loads and unloads

码头工人：在船坞上装卸车辆

English	Chinese
vehicles and handles freight on the dock.	处理货物的人。
Documents Against Acceptance: Instructions given by a shipper to a bank indicating that documents transferring title to goods should be delivered to the buyer (or drawee) only upon the buyer's payment of the attached draft.	承兑交单：由托运人给银行指示，用文件表明转移货物所有权，只有当买方的付款后才可以交付给买方。
Dog: Truck with little power.	"小狗"：小功率卡车。
Dolly: An auxiliary axle assembly having a fifth wheel used for purpose of converting a semitrailer to a full trailer.	拖带设备：一个辅助轴装配有第五轮，目的是改建半挂车为全挂车。
Domestic: Produced in the United States, including the Outer Continental Shelf (OCS).	国产：美国（包括外大陆架）生产的。
Domestic Fleet: All reportable agency owned motor vehicles, operated in any State, Commonwealth, Territory or possession of the United States.	国内车队：所有申报机构的国有机动车辆，在联邦地区或美国拥有的岛屿运营。
Domestic Intercity Trucking: Trucking operations within the territory of the United States, including intra-Hawaiian and intra-Alaskan, which carry freight beyond the local areas and commercial zones.	国内城际货运：美国领土内的货运业务，其中包括夏威夷和阿拉斯加，跨越地区和商业带进行货运。
Domestic Transportation: Transportation between places within the United States other than through a foreign country.	国内运输：在美国以内而不是运到其他国家的运输。
Donut Area: The area outside the Federal Highway Administration (FHWA) approved adjusted boundary of one or more urbanized	甜甜圈地区：一个处于联邦公路管理局规定的城市化地区以外，但在国家环境空气质量标

English	Chinese
areas but within the boundary of a National Ambient Air Quality Standards (NAAQ) non-attainment area.	准非到达领域的边界内的地区。
Donuts: Truck tires.	**甜甜圈**：卡车轮胎
Doodle Bug: Small tractor used to pull two axle dollies in a warehouse.	**室内小拖车**：用于仓库内作业的小型拖拉机拉两轴车。
Door Sill Step: Any step normally protected from the elements by the cab door when closed.	**门槛脚踏板**：任何在驾驶室车门关闭后被保护的脚踏板。
Double: (See also Rocky Mountain Double; Turnpike Double) A combination of two trailers pulled by a power unit. Usually refers to a power unit pulling two 28 foot trailers.	**双挂车组**：（又见洛基山双挂。收费公路双挂）由一个机组牵引的两部拖车的组合。通常是指一个牵引车拖动的两个28英尺的挂车。
Double Bottom: Unit consisting of tractor, semitrailer and full trailer. Also called "twin trailers", "doubles."	**双底拖拉机**：拖拉机，半挂车和全挂车的组合。也被称为"双拖车"。
Double Lockage: See Double.	**双打"**：见"双半挂车组"。
Down In the Corner: "Creeper" gear.	**下角**：爬行档
Downtime: The amount of time a vehicle or equipment is out of service for repair.	**停工时间**：车辆或设备进行修理或更换所需的时间。
Drawbridge: A bridge that pivots or lifts so as to let a boat through.	**可开闭吊桥**：桥梁可旋转或提升，以让船通过。
Driveaway-Towaway: Refers to a carrier operation, such as a fleet of tow trucks, used to transport other vehicles, when some or all wheels of the vehicles being transported touch the road surface.	**机动运载车**：是指承运人运行，如一个车队、拖车，用于运输其他车辆，部分或全部车轮接触路面。

English	Chinese
Driveaway-Towaway Operation: Any operation in which a motor vehicle constitutes the commodity being transported and one or more set of wheels of the vehicle being transported are on the surface of the roadway during transportation.	**机动运载**：对于商品运输机动车辆的任何操作。一套或多套轮子在运输过程中在地面上行驶。
Driver: See also Operator.	**司机**：见操作者。
Driver: 1) A person who operates a motorized vehicle. If more than one person drives on a single trip, the person who drives the most miles is classified as the principal driver. 2) An occupant of a vehicle who is in physical control of a motor vehicle in transport or, for an out of-control vehicle, an occupant who was in control until control was lost.	**司机**：1）凡经营机动车辆，如果一趟旅程不止一个人，驾驶里程最多的人为主要驾驶者 2）车辆里在运输过程中主要操控车辆的人。
Driver Applicant: An individual who applies to a State to obtain, transfer, upgrade, or renew a commercial driver's license (CDL).	**商业司机申请人**：向州政府申请取得、转换、升级或更换商业驾照的人。
Drivers: Drive wheels.	**驱动器**：驱动车轮。
Driver's License: A license issued by a State or other jurisdiction, to an individual which authorizes the individual to operate a motor vehicle on the highways.	**驾照**：一个州或其他管辖部门颁发给个人的许可证，向个人授权在公路驾驶车辆。
Driving a Commercial Motor Vehicle While Under the Influence of Alcohol: Committing any one or more of the following acts in a commercial motor vehicle (CMV): driving a CMV while the person's alcohol concentration is 0.04 percent or more; driv-	**酒后驾驶商业车辆**：任何一个或一个以上的下列行为：在酒精浓度为 0.04 ％或者更多时，在州政府法律界定的受酒精影响下时，驾驶商业机动车辆，或拒绝接受任何州或司法管辖

English	Chinese
ing under the influence of alcohol, as pre-scribed by State law; or refusal to undergo such testing as is required by any State or jurisdiction in the enforcement of 49 CFR 383.51b)2)(i)a) or b), or 49 CFR 392.5a)2).	规定的该项测试。
Driving Under the Influence (DUI): The driving or operating of any vehicle or common carrier while drunk or under the influence of liquor or narcotics.	**酒后驾车**：在喝醉酒或受酒或毒品影响下驾驶或操作任何车辆。
Drop the Body: Unhook and drive a tractor away from a parked semi.	**脱挂**：解开和驾驶拖拉机离开停车栏。
Dual Drive: (See also Tandem) Box axles have drive mechanisms and are connected to engine power output. 1) pusher tandem: only the rearmost axle is driving type and the forward unit is free rolling, also called "dead axle"; 2) tag axle: forward unit of tandem is driving type while rear unit is free rolling.	**双后桥布置**：（亦见串联）框车轴有驱动机制和连接发动机的动力输出。1）推串联：只有后轴驱动前部自由滚动，又称为"死桥"；2）标签车轴：前进股串联驱动型，而后方单位自由滚动。
Duals: A pair of tires mounted together.	**双轮胎**：安装在一起的一对轮胎。
Dump Body: Truck body of any type which can be tilted to discharge its load.	**侧卸式**：卡车上可倾斜的部分，用于卸货。
Dusting: Driving with wheels on road shoulder, thereby causing a cloud of dust.	**制造尘土**：车轮在道肩上驾驶造成的尘土飞扬。
Duty: A tax imposed by a government on imports.	**关税**：政府对进口商品所施加的关税。
Dynamic Routing; In demand-response transportation systems, the process of constantly modifying vehicle routes to	**动态路线选择**：在需求对应运输系统中，车辆开始运作后，不断修改车辆路线以适应收到

English	Chinese
accommodate service requests received after the vehicle began operations, as distinguished from predetermined routes assigned to a vehicle.	的不同的服务请求的过程。这有别于将货车分配到预定路线的过程。
Dynamite the Brakes: Emergency stop using every brake on the unit.	**紧急停车**：使用所有刹车单位来紧急刹车。

English	Chinese
Economy: Transport service established for the carriage of passengers at fares and quality of service below that of coach service.	**经济（舱）：**一般指票价和服务质量低于普通舱乘客运输服务。
Effective Locking Device: A manually operated switch or derail which is 1) Vandal resistant; 2) Tamper resistant; and 3) Capable of being locked and unlocked only by the class, craft or group of employees for whom the protection is being provided.	**有效锁定装置:** (用于保护铁道工作人员的)手动操纵的道岔或脱轨装置,具有以下特点 , 1)防肆意破坏; 2) 抗干预; 3)只能被保护的专职人员锁定和解锁。
Ejection: Refers to occupants being totally or partially thrown from the vehicle as a result of an impact or rollover.	**弹射:** 由于车辆碰撞或翻滚以致乘客整体或部分被抛出车外。
Elevated on Structure: Rail transit way above surface level on structure. Transition segments above surface level on structures are included.	**高架结构：**铁路高于路面的结构。包括高于地面的过渡结构。
Embankment: A raised structure of earth, ground, etc.	**高架结构:** (铁路运输) 指地面以上的高架铁路线. 也包括过渡段。
Emergency: A deviation from normal operation, a structural failure, or severe environmental conditions that probably would cause harm to people or property.	**紧急情况：**和正常运营不同 ,结构损害 , 或是严重的环境问题等 , 可能导致人员伤亡或财产损失。
Emergency: Any hurricane, tornado, storm (e.g. thunderstorm, snowstorm, icestorm, blizzard, sandstorm, etc.), high water, wind-driven water, tidal wave, tsunami, earth-quake, volcanic eruption, mud slide, drought, forest fire, explosion, blackout or other	**紧急情况:** 由自然或人为灾害 , 如飓风,龙卷风,暴风 , 高水位 , 海浪,海啸 地震,火山爆发,泥滑,干旱,森林火灾所造成的基本服务终端。 例如电力 , 医药服务 , 给排水 , 通讯等

English	Chinese
occurrence, natural or man-made, which interrupts the delivery of essential services (such as, electricity, medical care, sewer, water, telecommunications, and telecommunication transmissions) or essential supplies (such as, food and fuel) or otherwise immediately threatens human life or public welfare, provided such hurricane, tornado, or other event results in: a) A declaration of an emergency by the President of the United States, the Governor of a State, or their authorized representatives having authority to declare emergencies; by the Regional Director of Motor Carriers for the region in which the occurrence happens; or by other Federal, State or local government officials having authority to declare emergencies, or b) A request by a police officer for tow trucks to move wrecked or disabled vehicles.	或者食物和燃料。 另外， 灾害也可能对生命和财产直接的威胁。 有时飓风， 龙卷风， 或其他时件致使美国总统， 州长， 或他们的委托人宣布紧急状态。 另外， 当警察要求拖车把撞毁的车辆拖走时， 也叫紧急情况。
Emergency Brake: A mechanism designed to stop a motor vehicle after a failure of the service brake system.	**紧急制动闸**: 一个在正常刹车失效后使车辆停止的机械装置。
Emergency Brake System: A mechanism designed to stop a vehicle after a single failure occurs in the service brake system of a part designed to contain compressed air or brake fluid or vacuum (except failure of a common valve, manifold brake fluid housing or brake chamber housing).	**紧急制动系统**: 在正常车辆刹车系统失效后（通常是空气、液体或真空制动），使车辆停止的系统。但是不包括常用电子管、多种刹车液或是制动膛的失效。
Emergency Opening Window: That segment of a side facing glazing location	**紧急开放窗**：车辆侧窗，在紧急情况下可迅速并容易地拆

English	Chinese

which has been designed to permit rapid and easy removal during a crisis situation.

除。

Emergency Relief: An operation in which a motor carrier or driver of a commercial motor vehicle is providing direct assistance to supplement State and local efforts and capabilities to save lives or property or to protect public health and safety as a result of an emergency.

紧急援助: 在紧急情况下，卡车公司或卡车司机直接给政府提供援助以挽救生命或财产、保护公众健康和安全。

Emission Standards: Standards for the levels of pollutants emitted from automobiles and trucks. Congress established the first standards in the Clean Air Act of 1963. Currently, standards are set for four vehicle classes - automobiles, light trucks, heavy duty gasoline trucks, and heavy-duty diesel trucks.

排放标准: 汽车和卡车排放污染物的标准。美国国会在"1963 清洁空气法规"中制订了第一套标准。目前的标准为如下四类车辆设置: 汽车, 轻型卡车, 重型汽油(卡)车,重型柴油(卡)车。

Employee: 1) A driver of a commercial motor vehicle (including an independent contractor while in the course of operating a commercial motor vehicle); 2) A mechanic; 3) A freight handler; 4) Any individual, other than an employee, who is employed by an employer and who in the course of his or her employment directly affects commercial motor vehicle safety, but such term does not include an employee of the United States, any State, any political subdivision of a State, or any agency established under a compact between States and approved by the Congress of the United States who is acting within the course of such employment.

运输业雇员: 1) 受雇于经营商业运输卡车公司的司机， 2) 修理工， 3) 货物搬运工， 4) 任何员工, 在雇主雇佣的过程中, 他或她的工作直接影响到商业汽车/车辆的安全。但是这个名词不包括美国政府、州政府根据政府合约设立，由国会批准的代理机构的职员。

English	Chinese

Employee Hours: The number of hours worked by all employees of the railroad during the previous calendar year.

雇员工作时: 所有铁路雇员在过去一年里的工作小时总数。

Employee Human Factor: (See also Human Factor) Includes any of the accident causes signified by the rail equipment accident/incident cause codes listed under "Train Operation-Human Factors" in the current "Federal Railroad Administration (FRA) Guide for Preparing Accident/Incident Reports," except for Cause Code 506.

雇员人为因素: 指引起故障或事故的人为因素。事故原因包含现在的"联邦铁路管理委员会事故/事件预防指南"中的"车辆驾驶-人为因素"列表。但不包括代码 506。

Employer: Any person engaged in a business affecting interstate commerce who owns or leases a commercial motor vehicle in connection with that business, or assigns employees to operate it, but such terms does not include the United States, any State, any political subdivision of a State, or an agency established under a compact between States approved by the Congress of the United States.

雇主: 从事运输商务，拥有或租赁商用（机）车辆，并雇用员工驾驶，参与州际商业活动的经营人。但是这个名词不包括美国政府、州政府根据政府合约设立，由国会批准的代理机构。

End-Use Energy Consumption: Primary end-use energy consumption is the sum of fossil fuel consumption by the four end-use sectors (residential, commercial, industrial, and transportation) and generation of hydroelectric power by nonelectric utilities. Net end-use energy consumption includes electric utility sales to those sectors but excludes electrical system energy losses. Total end-use energy consumption includes both electric utility sales to the four

终端能源消费量: 四种终端用户 (住宅, 商业, 工业和交通运输业) 的总能源消费量，包扩来自矿物燃料和水力发电的能源。净终端用户能源消费量包括电力公司销售给终端用户的能源,但不包括电力系统的能耗损失。总终端能源总消费量包括电力公用事业销售给四种终端用户能源和电力系统的能耗

English	Chinese
end-use sectors and electrical system energy losses.	损失。
End-Use Sectors: The residential, commercial, industrial, and transportation sectors of the economy.	**终端能源用户部门**: 住宅、商业、工业和交通运输业。
Ending Milepost: The continuous milepost notation, to the nearest 0.01 mile that marks the end of any road or trail segment.	**终点英哩标示**: 道路终点英哩标示，最小标示距离为 0.01 英哩。
Endorsement: An authorization to an individual's commercial driver's license (CDL) required to permit the individual to operate certain types of commercial motor vehicles.	**签注**: 合格的商业驾照，持有人可操作该驾照规定的商用汽车。
Energy: The capacity for doing work as measured by the capability of doing work (potential energy) or the conversion of this capability to motion (kinetic energy). Energy has several forms, some of which are easily convertible and can be changed to another form useful for work. Most of the world's convertible energy comes from fossil fuels that are burned to produce heat that is then used as a transfer medium to mechanical or other means in order to accomplish tasks. Electrical energy is usually measured in kilowatt hours, while heat energy is usually measured in British thermal units.	**能量**: 衡量做功的能力(势能或动能)。能量有以下几种形式，其中一些容易转换，即从一种形式容易地转换成另外一种形式以便于使用。世界上大部分的能量来自矿物燃料燃烧时产生的热能，热是作为传递介质，以机械或其他形式来完成任务。 电能通常以千瓦小时来衡量，热能以英国热量单位来衡量。
Energy Consumption: The use of energy as a source of heat or power or as an input in the manufacturing process.	**能源消费**: 能源的使用,作为热源或动力源,或作为生产过程中的投入。

English	Chinese

Energy Efficiency: In reference to transportation, the inverse of energy intensiveness: the ratio of outputs from a process to the energy inputs; for example, miles traveled per gallon of fuel (mpg).

能源效率: 针对运输, 逆能源密集型: 能源投入和输出的比; 例如, 汽车每加仑燃油所走过英里数 (mpg)。

Energy Efficient Motors: Are also known as "high-efficiency motors" and "premium motors." They are virtually interchangeable with standard motors, but differences in construction make them more energy efficient.

节能发动机: 又称"高效发动机"和"高档马达", 它们几乎与普通发动机可以通用, 但不同的制造工艺, 使其更有效率地使用能源。

Energy Information Administration (EIA): An independent agency within the U.S. Department of Energy that develops surveys, collects energy data, and analyzes and models energy issues. The Agency must meet the requests of Congress, other elements within the Department of Energy, Federal Energy Regulatory Commission, the Executive Branch, its own independent needs, and assist the general public, or other interest groups, without taking a policy position.

能源信息管理署: 隶属于美国能源部的一个独立的机构, 主要职责包括调查, 收集能源数据, 建立模式和分析能源问题. 该机构必须满足国会其它能源部属下机构、联邦能源管理委员会、行政部门和自身的要求, 并协助该国一般公众或其他利益团体, 但不涉及政策立场。

Energy Intensity: In reference to transportation, the ratio of energy inputs to a process to the useful outputs form that process; for example, gallons of fuel per passenger-mile or BTU per ton-mile.

能源强度:(运输业) 能量输入和经过能量转换过程后输出的有益产功力之比, 例如, 每加仑客英里或每 BTU 吨英里。

Energy Source: A substance, such as petroleum, natural gas, or coal that supplies heat or power. In Energy Information Administration reports, electricity and renew-

能源: 一种产生能量的物质, 如石油,天然气和煤炭, 热或电力。根据能源信息管理署报告, 电力和可再生的能源,例如生

English	Chinese
able forms of energy, such as biomass, geothermal, wind, and solar, are considered to be energy sources.	物, 地热,风力和太阳能 被认为是能源。
Engine Classification: A 2-digit numeric code identifying vehicle engines by the number of cylinders.	**引擎分类:** 根据发动机气缸数量指定的两位数码系统。
Engine Displacement: (See also Cylinder, Engine Size) The volume in inches, through which the head of the piston moves, multiplied by the number of cylinders in the engine. Also known as cubic inch displacement (CID), may also be measured in liters.	**引擎排量:**(参见气缸，发动机大小)通过活塞气缸的大小再乘以气缸的数量来衡量，以立方英寸（或升）为单位。
Engine Retarder: Electronic equipment which governs engine speed control.	**发动机缓速器:** 用于调节发动机速度控制的电子装置。(用于调节发动机辅助制动的电子装置)。
Engine Size: (See also Cylinder, Engine Displacement) The total volume within all cylinders of an engine, when pistons are at their lowest positions. The engine is usually measured in "liters" or "cubic inches of displacement (CID)." Generally, larger engines result in greater engine power, but less fuel efficiency. There are 61.024 cubic inches in a liter.	**发动机大小:**(参见气缸，引擎排量) 当活塞处于最低位置时气缸的总容量。通常以"立升"或"立方英寸（CID）"来衡量。一般而言，较大的引擎会产生更大的发动机功率,但较低燃油效率。61.024 立方英寸等于一公升。
Entrapment: Refers to persons being partially or completely in the vehicle and mechanically restrained by a damaged vehicle component. Jammed doors and immobilizing injuries, by themselves, do not constitute entrapment.	**受困:** 指人（们）因受损车辆部分或完全受困于车内的情况，不包括因车门失灵、自我受伤、、车载货物移动受压、撞击被抛出而受困，或安全带失灵而造成的情况。

English	Chinese

Environmental Protection Agency Certification Files: Computer files produced by Environmental Protection Agency (EPA) for analysis purposes. For each vehicle make, model and year, the files contain the EPA test Miles Per Gallon (MPG) (city, highway and 55/45 composite). These MPG's are associated with various combinations of engine and drive-train technologies (e.g., number of cylinders, engine size, gasoline or diesel fuel, and automatic or manual transmission).

环境保护局认证档案: 环境保护局 (EPA) 建立的用于分析的电子档案。该档案包含环保局检验每个汽车品牌,型号和年份的每加仑英里(mpg)(例如高速公路和市区组合 mpg 为 55/45)。这些 mpg 是基于各种因素的组合(例如,气缸数,引擎的大小, 汽油或柴油,以及自动或手动传送等)。这些档案还载有在能源部(DOE)/环保局气哩指南类似的资料。

Environmental Protection Agency Composite Mile Per Gallon (MPG): The harmonic mean of the Environmental Protection Agency (EPA) city and highway MPG, weighted under the assumption of 55 percent city driving and 45 percent highway driving.

(美国)环保局综合每加仑英里 (MPG): 环境保护局(EPA)定义的城市和公路的调和平均 MPG, 是基于 55%的城市驾驶和 45%的公路驾驶的加权平均。

Equalizing Reservoir: An air reservoir connected with and adding volume to the top portion of the equalizing piston chamber of the automatic brake valve, to provide uniform service reductions in brake pipe pressure regardless of the length of the train.

均衡风缸: 风缸与自动刹车阀的均一活塞相连 , 可添加流量到其顶端部分。不论列车长度 , 提供统一的服务,减少刹车油管压力。

Equipment Code: A six digit numeric code used to classify equipment by its usage characteristics (passenger carrying, cargo hauling, etc.), gross weight rating, and equipment configuration (panel truck, pick-up, stake body, dump etc.).

设备代码:用来区分设备的使用特性(载客,货物托运, 等), 毛重评级及设备配置(厢式货车,皮卡,支体,泥等)的六位数字系统。

English	Chinese
Event: See also Accident, Casualty, Collision, Crash, Derailment, Fatality, Incident, and Injury.	**事件：** 参照事故。如人员伤亡、相撞、坠毁、脱轨、死亡受伤等意外事故。
Event Recorder: A device, designed to resist tampering, that monitors and records data on train speed, direction of motion, time, distance, throttle position, brake applications and operations (including train brake, independent brake, and, if so equipped, dynamic brake applications and operations) and, where the locomotive is so equipped, cab signal aspect(s), over the most recent 48 hours of operation of the electrical system of the locomotive on which it is installed.	**事件纪录器：** 能防范人为篡改，监测和记录最近的 48 小时有关电器设备启用后的车速，运行动方向，时间，距离，油门位置，刹车使用，以及驾驶室信号显示状态。
Exclusion Zone: An area surrounding a Liquefied Natural Gas (LNG) facility in which an operator or government agency legally controls all activities in accordance with 49 CFR 193.2057 and 49 CFR 193.2059 for as long as the facility is in operation.	**禁区：** 由政府或私人企业管控的液化天然气设施区域。
Exempt Carrier: A for hire interstate operator [which] transports commodities or provides types of services that are exempt from federal regulation, could also operate within exempt commercial zones.	**非控管承运人：** 不受联邦运输经济法管制的，在非控管商业区操作的从事州际商业运输业务的承运人。
Exempt Intracity Zone: The geographic area of a municipality or the commercial zone of that municipality described by the Interstate Commerce Commission (ICC) in 49 CFR 1048, revised as of October 1, 1975.	**豁免城区：** 由州际商业署划定的市区或商业区。 这些区划在 1975 年十月一日更新过。

English	Chinese

Exempt Motor Carrier: A person engaged in transportation exempt from economic regulation by the Interstate Commerce Commission (ICC) under 49 U.S.C. 10526.

豁免货车公司：由州际商业署批准，免受联邦运输经济法约束的从事运输业的货车公司。

Exemption: A temporary or permanent grant, license or form of legal permission given by an agency to deviate from a regulation or provision of law administered by that agency. Issued in response to a petition for relief submitted by an individual or company.

特许证：由政府机构签发的合法的临时或永久许可证。此许可证不同于政府机构按常规发放的许可证，专为赦免某些个人或公司申诉的特殊情况而签发。

Expandable: Flatbed trailer which can be expanded beyond its regular length to carry larger shipments.

可延伸板车：可延伸的平板车，用于装载大型货物。

Express Body: Open box truck body.

敞箱式车身：无盖卡车车厢。

Expressway: (See also Freeway, Freeways and Expressways, Highway, Interstate Highway (Freeway or Expressway), Road). A divided highway for through traffic with full or partial access control and including grade separations at all or most major intersections.

高速公路：双向隔开，有进出管制，所有或多数交叉路口是立体交叉的公路。

Extraordinary Items Income (Loss): Income or loss which can be characterized as material, unusual and of infrequent occurrence.

非正常收入（损失）：确实的，非正常、不经常发生的财物收入或损失。

English	Chinese
Facility: All or any portion of buildings, structures, sites, complexes, equipment, roads, walks, passageways, parking lots, or other real or personal property, including the site where the building, property, structure, or equipment is located.	**设施**：建筑、结构、场所、综合建筑群、设备，道路、人行道，过道，停车场，或者其它不动产以及个人所有物。包括上述内容所位于的场所。
Fair Market Value: The value of a vehicle as stated by the National Automotive Dealers Association (NADA) or other sale publication. For vehicles under the 3-year replacement cycle, Fair Market Value is the average loan indicated in the appropriate NADA publication.	**车辆公平市价**：由全美汽车商协会（NADA）或者其它销售出版物确定的车辆价值。对少于三年更新周期的车辆，公平市价为相关NADA出版物中所指示的平均贷款额。
Farm-To-Market Agricultural Transportation: The operation of a motor vehicle controlled and operated by a farmer who: 1) Is a private motor carrier of property; 2) Is using the vehicle to transport agricultural products from a farm owned by the farmer, or to transport farm machinery or farm supplies to or from a farm owned by the farmer; and 3) Is not using the vehicle to transport hazardous materials of a type or quantity that require the vehicle to be placarded in accordance with 49 CFR 177.823.	**农场至市场农业运输**：由农场主控制并运营的机动车辆运输。该农场主是：1）机动车辆的私人财产拥有者；2）使用车辆运输自己农场的农产品、农场机械、农场供给；3）不使用车辆运输根据49 CFR 177.823必须对车辆加以告示说明的特定种类或者特定数量的危险物质。
Farm Vehicle Driver: A person who drives only a motor vehicle that is 1) Controlled and operated by a farmer as a private motor carrier of property; 2) Being used to transport either agricultural products, or farm machinery, farm supplies, or both, to or from	**农用车驾驶员**：只能驾驶以下机动车辆的人员：1）机动车辆为农场主的私有财产并由其运营；2）使用车辆从自己的农场运输农产品，或者运输农场机械、农场供给；3）车辆

English	Chinese
a farm; 3) Not being used in the operation of a for-hire motor carrier; 4) Not carrying hazardous materials of a type or quantity that requires the vehicle to be placarded in accordance with 49 CFR 177.823 and 5) Being used within 150 air-miles of the farmer's farm.	不用于出租；4）不使用车辆运输根据 49 CFR 177.823 必须告示说明的特定种类或者特定数量的危险物质；5）活动范围限于农场主农场 150 空英里范围内。
Fatal Accident: 1) A motor vehicle traffic accident resulting in one or more fatal injuries. 2) An accident for which at least one fatality was reported.	**死亡事故**：1）导致一人或多人致命伤害的机动车辆交通事故。2）至少一人死亡的事故。
Fatal Accident: (See also Fatality) Statistics reported to the Federal Highway Administration (FHWA) shall conform to the 30-day rule, i.e., a fatality resulting from a highway vehicular accident is to be counted only if death occurs within 30 days of the accident.	**死亡事故报告**：（参见死亡）报告给联邦公路管理局（FHWA）的统计报告，遵循 30 天准则，只统计公路车辆事故 30 天内导致的死亡。
Fatal Accident: An accident that results in one or more deaths within one year.	**死亡事故**：导致一人或多人在一年内死亡的事故。
Fatal Accident Rate: The fatal accident rate is the number of fatal accidents per 100 million vehicle miles of travel.	**死亡事故率**：每 1 亿车辆行驶里程死亡事故数量。
Fatal Alcohol Involvement Crash: A fatal crash is alcohol-related or alcohol-involved if either a driver or a non motorist (usually a pedestrian) had a measurable or estimated blood alcohol concentration (BAC) of 0.01 grams per deciliter (g/dl) or above.	**酒精性致命车辆碰撞**：如果测量或者估计驾驶员或者非机动车驾驶员（通常是行人）血液酒精浓度（BAC）达到或者超过 0.01 克每 0.1 公升(g/dl)，那么，该致命性车辆碰撞即为酒精性致命车辆碰撞。

English	Chinese
Fatal Crash: A police-reported crash involving a motor vehicle in transport on a trafficway in which at least one person dies within 30 days of the crash.	**致命碰撞**：警察局备案的涉及到机动车辆并且发生在公路上的致命性车辆碰撞事故，在该事故发生的三十天内，至少有1人因这起事故死亡。
Fatal Injury: Any injury which results in death within 30 days of the accident.	**致命性伤害**（2）：在事故发生后的三十天内导致死亡的任何伤害。
Fatal Plus Nonfatal Injury Accidents: The sum of all fatal accidents and nonfatal-injury accidents.	**致命性及非致命性伤害事故**：所有致命性和非致命性伤害事故之和。
Fatality: Are those 1) Which result from motor vehicle accidents that occurred during the relevant calendar year and 2) Those in which the injured person(s) died within 30 days of the accident.	**死亡**：指1）在历年内由相关车辆交通事故导致的死亡；2）事故发生三十天内受伤人员的死亡。
Fatality: A death as the result of a crash that involves a motor vehicle in transport on a trafficway and in which at least one person dies within 30 days of the crash.	**死亡**：在公路上机动车辆的碰撞事故发生三十天内引起的死亡。
Fatality: For purposes of statistical reporting on transportation safety fatality shall be considered a death due to injuries in a transportation accident or incident that occurs within 30 days of that accident or incident.	**死亡**：交通运输安全统计报告的死亡数据，即在交通事故或者事件发生的30天内由于在事故或事件中受伤而导致的死亡。
Fatality/Injury: Refers to the average number of fatalities and injuries which occurred per one hundred accidents. Frequently used as an index of accident severity.	**死伤人数**：指每一百起事故的平均死亡数和受伤人员数。经常被用作事故严重性指标。

English	Chinese
Fatality Rate: The average number of fatalities which occurred per accident or per one hundred accidents.	**死亡率**：每起事故或者每一百起事故的平均死亡数。
Fatality Rate: The fatality rate is the number of fatalities per 100 million vehicle miles of travel.	**死亡率**: 每 1 亿车辆行驶里程死亡数。
Federal-Aid Highways: Those highways eligible for assistance under Title 23 U.S.C. except those functionally classified as local or rural minor collectors.	**联邦资助公路**：除当地或者乡村次要连接点的公路外，在 23 U.S.C.条款下符合援助条件的公路。
Federal-Aid Primary Highway System: The Federal-Aid Highway System of rural arterials and their extensions into or through urban areas in existence on June 1, 1991, as described in 23 U.S.C. 103b in effect at that time.	**联邦资助主要公路系统**：在 1991 年 6 月 1 日生效的 23U.S.C. 103b) 中所规定的、由农村主干道及其向市区或经过市区的延伸部分所构成的联邦资助公路系统。
Federal Aid Secondary Highway System: This existed prior to the ISTEA [Intermodal Surface Transportation Efficiency Act] of 1991 and included rural collector routes.	**联邦资助二级公路系统**：该系统早于在 1991 年通过的 ISTEA（陆路联合运输效率法），包括农村联结线路。
Federal Aid Urban Highway System: This existed prior to the ISTEA [Intermodal Surface Transportation Efficiency Act] of 1991 and included urban arterial and collector routes, exclusive of urban extensions of the Federal-Aid Primary system.	**联邦资助城市公路系统**：该系统早于 1991 年通过的陆路联合运输效率法，包括城市主干道和连接线路，不包括联邦资助主要公路系统内的城市延伸区域。
Federal Highway Administration (FHWA): Became a component of the Department of Transportation in 1967 pursuant to the	**联邦公路管理局（FHWA）**：根据交通部法案(49 U.S.C. app. 1651 note)，于 1967 年

English

Department of Transportation Act (49 U.S.C. app. 1651 note). It administers the highway transportation programs of the Department of Transportation under pertinent legislation and the provisions of law cited in section 6a) of the act (49 U.S.C. 104) The Administration encompasses highway transportation in its broadest scope seeking to coordinate highways with other modes of transportation to achieve the most effective balance of transportation systems and facilities under cohesive Federal transportation policies pursuant to the act. The Administration administers the Federal-Aid Highway Program; is responsible for several highway-related safety programs; is authorized to establish and maintain a National Network for trucks; administers a coordinated Federal lands program; coordinates varied research, development and technology transfer activities; supports and participates in efforts to find research and technology abroad; plus a few additional programs.

Federal Motor Carrier Safety Regulations (FMCSR): The regulations are contained in the Code of Federal Regulations, Title 49, Chapter III, Subchapter B.

Federal Power Act: Enacted in 1920, and amended in 1935, the Act consists of three parts. The first part incorporated the Federal Water Power Act administered by the former Federal Power Commission, whose activities were confined almost entirely to licensing

Chinese

成为交通部的一个组成部分。它在法案（（49 U.S.C. 104））的 6a）节内所规定的相关法律条款范围内管理公路运输项目。根据该法案及相关联邦运输政策，为了达到运输系统和设备的最有效的平衡，联邦公路管理局最大限度地将公路运输与其它运输模式相协调。联邦公路管理局管理联邦资助公路项目、负责与公路有关的安全项目、被授权建立和维护全国卡车网络、管理联邦土地协调项目、协调各种研究、开发和技术转让活动，并支持和参与在国外的研究和技术开发，另外还有一些附加项目。

联邦汽车运输安全规程 (FMCSR)：该规程包含在联邦法规汇编的第 III 章，B 次章，第 49 款中。

联邦能源法案：该法案颁布于 1920 年，于 1935 年进行修正，包括三部分。第一部分整合了前联邦能量委员会管理执行的联邦水能源法案。该委员会的主要活动主要是向非联邦

English	Chinese

non-Federal hydroelectric projects. Parts II and III were added with the passage of the Public Utility Act. These parts extended the Act's jurisdiction to include regulating the interstate transmission of electrical energy and rates for its sale as wholesale in interstate commerce. The Federal Energy Regulatory Commission is now charged with the administration of this law.

水力电气工程颁发许可证。第二部分和第三部分加上了公用事业法案。这些部分扩充了该法案的权限，包括电力能源州间传输的管制，州间电力批发销售定价。联邦能源管理委员会现在负责该法律的管理执行。

Federal Register: Daily publication which provides a uniform system for making regulations and legal notices issued by the Executive Branch and various departments of the Federal government available to the public.

联邦公报：向公众提供由行政部门和联邦政府各部门颁布的法规和法律通告的系列日常出版物。

Federal Water Pollution Control Act (FWPCA): Law passed in 1970 and amended in 1972 giving the Coast Guard a mandate to develop, among other things, marine sanitation device regulations.

联邦水污染控制法案（FWPCA）：该法案于 1970 年通过并于 1972 年修正。它给予海岸警备队的要求之一就是制定海运卫生设施法规。

Fifth Wheel: A device mounted on a truck tractor or similar towing vehicle (e.g., converter dolly) which interfaces with and couples to the upper coupler assembly of a semitrailer.

转向轮：用来支撑安装在车体结构上的钢板的装置。它固定于轮轴之上，可以支撑半拖挂车的中心轴，在牵引机和拖车之间提供灵活的连接。

Finished Gasohol Motor Gasoline: (See also Gasohol, Gasoline) A blend of finished motor gasoline (leaded or unleaded) and alcohol (generally ethanol, but sometimes methanol) in which 10 percent or more of the product is alcohol.

汽醇发动机汽油：（参见汽醇，汽油）发动机汽油（含铅或不含铅）和酒精（通常是乙醇，有时是甲醇）的混合物，其中，酒精占百分之十或更多。

English	Chinese
Finished Gasoline: See also Gasoline.	**成品汽油**：参见汽油。
Finished Leaded Gasoline: Contains more than 0.05 gram of lead per gallon or more than 0.005 gram of phosphorus per gallon. Premium and regular grades are included, depending on the octane rating. Includes leaded gasohol. Blendstock is excluded until blending has been completed. Alcohol that is to be used in the blending of gasohol is also excluded.	**成品含铅汽油**：每加仑汽油含铅量超过 0.05 克或者含磷量超过 0.005 克。取决于辛烷含量，有高级汽油和普通汽油之分。包括含铅汽醇。除非混合过程已经完成，不包括未混合的氧合汽油。也不包括汽醇混合的酒精。
Finished Leaded Premium Motor Gasoline: Motor gasoline having an antiknock index, calculated as (R+M)/2, greater than 90 and containing more than 0.05 gram of lead per gallon or more than 0.005 gram of phosphorus per gallon.	**成品含铅高级动力汽油**：抗爆指数大于 90 的动力汽油。计算式为(R+M)/2。每加仑含铅量超过 0.05 克或者含磷量超过 0.005 克。
Finished Leaded Regular Motor Gasoline: Motor gasoline having an antiknock index, calculated as (R+M)/2, greater than or equal to 87 and less than or equal to 90 and containing more than 0.05 gram of lead or 0.005 gram of phosphorus per gallon.	**成品含铅普通动力汽油**：抗爆指数大于或等于 87 且小于或等于 90 的动力汽油，计算式为(R+M)/2，每加仑含铅量超过 0.05 克或者含磷量超过 0.005 克。
Finished Motor Gasoline: 1) A complex mixture of relatively volatile hydrocarbons, with or without small quantities of additives, blended to form a fuel suitable for use in spark-ignition engines. Specification for motor gasoline, as given in American Society for Testing and Materials (ASTM) Specification D439-88 or Federal Specification VV-G-1690B, include a boiling range of 122 degrees to 158 degrees Fahrenheit at the	**成品动力汽油**：1）含有少量添加剂或不含有添加剂的相对挥发性碳氢复杂混合物，经混合后形成适用于火花点燃式发动机的燃料。在美国材料试验协会（ASTM）D439-88 规范或者美国联邦标准 VV-G-1690B 中，动力汽油的范围在 365 度的 10%分位点和 374 度的 90%分位点，沸点范围为

English	Chinese
10-percent point to 365 degrees to 374 degrees Fahrenheit at the 90-percent point and a Reid vapor pressure range from 9 to 15 psi. "Motor gasoline" includes finished leaded gasoline, finished unleaded gasoline, and gasohol. Blendstock is excluded until blending has been completed. (Alcohol that is to be used in the blending of gasohol is also excluded.) 2) Motor gasoline that is not included in the reformulated or oxygenated categories.	122 至 374 华氏摄氏度，里德蒸气压范围为 9 至 15 psi。"动力汽油"包括含铅汽油，不含铅汽油和汽醇，以及未经混合的氧合汽油。（也不包括汽醇混合的酒精。）2）动力汽油不包含在新配方汽油或者氧合汽油的类别内。
Finished Unleaded Gasoline: Contains not more than 0.05 gram of lead per gallon and not more than 0.005 gram of phosphorus per gallon. Premium and regular grades are included, depending on the octane rating. Includes unleaded gasohol. Blendstock is excluded until blending has been completed. Alcohol that is to be used in the blending of gasohol is also excluded.	**成品不含铅汽油**：每加仑含铅量不超过 0.05 克或者含磷量不超过 0.005 克。取决于辛烷含量，有高级汽油和普通汽油之分。包括含铅汽醇。不包括未经混合的氧合汽油和用于汽醇混合的酒精。
Finished Unleaded Midgrade Motor Gasoline: Motor gasoline having an antiknock index, calculated as (R+M)/2, greater than or equal to 88 and less than or equal to 90 and containing not more than 0.05 gram of phosphorus per gallon.	**成品不含铅中等动力汽油**：抗爆指数大于或等于 88 且小于或等于 90 的动力汽油，计算式为(R+M)/2。每加仑含铅量不超过 0.05 克或者含磷量不超过 0.005 克。
Finished Unleaded Premium Motor Gasoline: Motor gasoline having an antiknock index, calculated as (R+M)/2, greater than 90 and containing not more than 0.05 gram of lead or 0.005 gram of phosphorus per gallon.	**成品不含铅高级动力汽油**：抗爆指数大于 90 的动力汽油，计算式为(R+M)/2。每加仑含铅量不超过 0.05 克而且含磷量不超过 0.005 克。

English	Chinese
Finished Unleaded Regular Motor Gasoline: Motor gasoline having an antiknock index, calculated as (R+M)/2, of 87 containing not more than 0.05 gram of lead per gallon and not more than 0.005 gram of phosphorus per gallon.	**成品不含铅普通动力汽油**：抗爆指数为 87 的动力汽油，计算式为(R+M)/2。每加仑含铅量不超过 0.05 克而且含磷量不超过 0.005 克。
Fire/Explosion, Fuel: Accidental combustion of vessel fuel, liquids, including their vapors, or other substances, such as wood or coal.	**失火或者爆裂（参见防爆燃性）**：由轨道交通承载输送的材料突然燃烧或者猛烈释放而导致的事故/事件。这类事故包括：燃料和电力设备起火、曲轴箱爆炸、液化石油或者无水氨猛烈释放。
First Harmful Event: A first harmful event is the first event during a traffic accident that causes an injury (fatal or nonfatal) or property damage.	**最初有害事件**：指在交通事故中最先导致受害（致命性或非致命性）或者财产损失的事件。
Fixed Collision Barrier: A flat, vertical, unyielding surface with the following characteristics: 1) The surface is sufficiently large that when struck by a tested vehicle, no portion of the vehicle projects or passes beyond the surface; 2) The approach is a horizontal surface that is large enough for the vehicle to attain a stable attitude during its approach to the barrier, and that does not restrict vehicle motion during impact; 3) When struck by a vehicle, the surface and its supporting structure absorb no significant portion of the vehicle's kinetic energy, so that a performance requirement described in terms of impact with a fixed collision barrier	**固定防撞栏**：具有如下特征的平坦、直立并且坚硬的面状结构：1）表面足够大，当测试车辆撞击时，车辆没有任何部分射出或者越过该表面；2）接触面是一个足够大的水平面，当车辆接近防撞栏时可以得到一个稳定的接触面，并且在冲突过程中不会限制车辆运动；3）当受到车辆撞击时，其表面和支撑结构吸收车辆的动能有限，因此固定防撞栏的性能表现必须可靠，无论防撞栏需要吸收的能量有多小。

English	Chinese

must be met no matter how small an amount of energy is absorbed by the barrier.

Fixed Crane: A crane whose principal structure is mounted on a permanent or semipermanent foundation.

固定式起重机：其基本结构固定在一个永久性或半永久性基础上的起重机。

Fixed Object: Stationary structures or substantial vegetation attached to the terrain.

固定对象：固定结构或者依附于地表的大量植被。

Fixed Operating Cost: In reference to passenger car operating cost, refers to those expenditures that are independent of the amount of use of the car, such as insurance costs, fees for license and registration, depreciation and finance charges.

固定运营成本：与客车运营成本有关，指与车辆的使用次数无关的花费，比如保险费、执照申请费、注册费、折旧及财务费用。

Fixed Route: Service provided on a repetitive, fixed-schedule basis along a specific route with vehicles stopping to pick up and deliver passengers to specific locations; each fixed-route trip serves the same origins and destinations, unlike demand response and taxicabs.

固定路线：基于重复、固定的时刻表，沿着某一具体路线，车辆停车载客，将乘客送往具体目的地的一种服务。与需求响应式及出租车不同，每一固定路线行程具有相同的起讫点。

Fixed Route System: A system of transporting individuals (other than by aircraft), including the provision of desig-nated public transportation service by public entities and the provision of transportation service by private entities, including, but not limited to, specified public transportation service, on which a vehicle is operated along a prescribed route according to a fixed schedule.

固定路线系统：个体运输系统（不同于航空运输），包括由政府部门指定的公共交通服务，由私营实体提供的交通服务，包括但不局限于车辆按照固定时刻表沿着事先规定的路线提供的特定公共交通服务。

English	Chinese

Flag Drop Charge: The charge for an initial distance (usually specified by regulation) for taxi service. It is actually the minimum fare.

起步价：出租车服务所收取的最短行程费用（通常由相关法规规定）。它实际上是最小费用。

Flagman's Signals: A red flag by day and a white light at night, and a specified number of torpedoes and fusees as prescribed in the railroad's operating rules.

旗员信号：白天为红旗，晚上为白光，在铁路操作规范中事先规定的一定数量的响墩信号和引信。

Flame Resistant: Not susceptible to combustion to the point of propagating a flame, beyond safe limits, after the ignition source is removed.

耐燃性：不易燃烧，当火源移除，不会导致火焰传递。

Flammable: With respect to a fluid or gas, means susceptible to igniting readily or to exploding.

可燃性：易着火或爆炸的物体，通常指液体或者气体。

Flash Resistant: (See also Fire or Violent Rupture) Not susceptible to burning violently when ignited.

防爆燃性（参见失火或者爆裂）当点燃时不容易剧烈燃烧。

Flasher: In rail systems, the flashing light at railroad grade crossings that warns motorists, bicyclists, and pedestrians of approaching trains.

闪灯：铁路系统和地面道路交叉口处的闪灯，用于提醒机动车，自行车和行人火车正在驶近。

Flat Bottom: Flatbed.

平板货车：平板载货车。

Flat Car: A rail car without a roof and walls.

铁路平板车：一种没有顶篷和车壁的铁路用车。

Flat Face: Cab over engine.

平面室：位于发动机上方的驾驶室。

English	Chinese

Flat Rate Manual: A manual published by an equipment manufacturer or an independent publisher that indicates the length of time required for performing specific mechanical tasks such as installing a clutch. Normally, the costs of parts required for a specific job are also listed.

按时计价手册：由设备制造商或者独立出版商出版的手册，用来说明完成具体的机械操作比如安装离合器所需要的时间量。通常，具体工作所需各项费用也被列出。

Flatbed: Truck or trailer without sides and top.

平底车：没有车壁和车顶的货车或拖车。

Fleet: The vehicles in a transit system. Usually, "fleet" refers to highway vehicles and "rolling stock" to rail vehicles.

车队：处于运输系统中的车辆。通常，"车队"指公路车辆，"机车车辆"指铁路车辆。

Fleet Management Center (FMC): A formally approved element of the Interagency Fleet Management System (IFMS) responsible for the administrative control of Interagency Fleet Management System (IFMS) vehicles in a specified geographic area as defined in the determination that is approved by the Administrator of General Services.

车队管理中心（FMC）：被正式认可的部门间车辆管理控制系统（IFMS）组成要素，负责对通用服务管理处划分的某一地区内的车队管理系统（IFMS）进行管理控制。

Fleet Management Subcenter: A formally approved element of the Interagency Fleet Management System (IFMS) Fleet Management Center physically detached from the central or main Fleet Management Center.

车队管理子中心：被联邦管理总署正式认可的部门间车辆管理系统（IFMS）组成要素，地点上与车队管理主中心分开。

Fleet Management System (FMS): The automated inventory and control system used by the Interagency Fleet Management System (IFMS) to track vehicle assignments,

车队管理系统（FMS）：部门间车队管理系统（IFMS）。是一个自动盘存和控制系统，用来跟踪车辆分配、车辆利用情

English	Chinese
vehicle utilization, and provide direct input to the Finance Division to bill customer agencies for the use of IFMS vehicles.	况并且向财政部门直接提供数据、为使用 IFMS 车辆的客户开出帐单。
Fleet Vehicles: 1) Private fleet vehicles: ideally, a vehicle could be classified as a member of a fleet if it is operated in mass by a corporation or institution, operated under unified control, or used for non-personal activities; however, the definition of a fleet is not consistent throughout the fleet industry. Some companies make a distinction between cars that were bought in bulk rather than singularly, or whether they are operated in bulk, as well as the minimum number of vessels that constitute a fleet (i.e. 4 or 10); 2) Government fleet vehicles: includes vehicles owned by all federal (GSA) state, county, city, and metro units of government, including toll road operations.	**车队车辆**：1）私有车队车辆：在理想情况下，如果车辆被公司或团体集体操作并且在统一控制下运营，或者被用作非个人用途，该车辆为车队车辆。然而，车队的定义在运输工业界并不统一。一些公司将车辆区分为批量购入车辆和单一购入车辆，或者按它们是不是同时运营及购成车队的最小车辆数（比如四辆或者十辆）来进行区分；2）政府车队：包括被所有联邦州、县、城市和地区政府，包括收费道路运营部门所拥有的车辆。
Float: Flatbed semitrailer.	**平板车**：平板半拖车。
Flyer: A run in which the driver takes a trailer to a distant terminal, leaves it there and immediately pulls another trailer back to his home terminal.	**往返拖运**：驾驶员将一台拖车拖到远点终端，把它留在那里，并立刻将另一台拖车拖回到起点终端的过程。
For Hire: Refers to a vehicle operated on behalf of or by a company that provides transport services to its customers.	**出租车**：指代表出租运营公司向顾客提供运输服务的的车辆。
For-Hire Carriage: Transportation of property by motor vehicle except when: 1) The property is transported by a person engaged in a business other than	**雇佣车**：运输财产的机动车辆，不包括以下车辆：1）运送者的主要业务不是交通；2）促进其它非运输活动的商

English	Chinese
transportation; and 2) The transportation is within the scope of, and furthers a primary business (other than transportation) of, the person.	业活动和为运送者个人服务的车辆。
For-Hire Motor Carrier: A person engaged in the transportation of goods or passengers for compensation.	**出租车司机**：运送货物或者乘客以获得报酬者。
Foreign Fleet: All reportable agency owned motor vehicles, operated outside any State, Commonwealth, Territory or possession of the United States.	**涉外车队**：在美国任何州、联邦、属地或者领土之外运营的、被组织所拥有的所有上报的机动车辆。
Foreign Freight: Movements between the United States and foreign countries and between Puerto Rico, the Virgin Islands, and foreign countries. Trade between U.S. territories and possessions (e.g. Guam, Wake, American Samoa) and foreign countries is excluded. Traffic to or from the Panama Canal Zone is included.	**境外运输**：在美国和其它国家以及在波多黎各、维尔京群岛和其它国家之间进行的运输。不包括美国国土和美属领地（比如关岛，威克岛和美属萨摩亚）内及外国之间的商业运输。包括驶向或者驶出巴拿马运河区的交通运输。
Foreign Freight Forwarder: An independent business which makes shipments for exporters for a fee.	**国外货运承揽者**：向出口商提供运输服务并收取费用的独立商业运输者。
Forestall: As applied to an automatic train stop or train control device, to prevent an automatic brake application by operation of an acknowledging device or by manual control of the speed of the train.	**阻止器**：指列车自动停止或控制设备，通过确认装置或者手动控制列车的速度来防止超速制动。
Forward Control: A configuration in which more than half of the engine length is rearward of the foremost point of the windshield base and the steering wheel hub	**前向控制**：超过一半发动机长度在挡风玻璃基部最前点的后面，并且方向盘在车长的前四

English	Chinese
is in the forward quarter of the vehicle length.	分之一处的结构配置。
Forward Control: Vehicle with driver controls (pedals, steering wheel, instruments) located as far forward as possible. Supplied with or without body, the controls are stationary mounted as opposed to the special mountings of tilt cabs.	**前向控制**：驾驶员控制设备（踏板、方向盘和仪表）位于尽可能最前端的车辆。不论具不具有单独形体，这些控制设备都与驾驶室的特殊装备相对固定安装。
Fossil Fuel: Any naturally occurring organic fuel, such as petroleum, coal, and natural gas.	**化石燃料**：任何天然形成的有机燃料，比如石油、煤和天然气。
Four Banger: Four cylinder engine.	**四气缸发动机**：带有四个汽缸的发动机。
Four By Four: Four speed transmission and 4 speed auxiliary transmission.	**四驱**：四级变速传动和四级辅助变速传动。
Frangible Navigational Aid (NAVAID): A navigational aid whose properties allow it to fail at a specified impact load.	**易碎导航设备（NAVAID）**：在一定的压力负载下会被损坏的导航设备。
Freeway: See also Arterial, Expressway, Highway, Interstate Highway (Freeway or Expressway), Local Streets and Roads, Road.	**高速公路**：参见主干道、快速路、公路、州际公路（高速公路或快速路）、街区道路和道路。
Freeway: An expressway with full control of access.	**高速公路**：入口完全控制的快速路。
Freeways and Expressways: (See also Expressway, Highway, Interstate Highway (Freeway or Expressway)) All urban principal arterial with limited control of access not on the interstate system.	**高速公路和快速路**：（参见快速路、公路、州际公路（高速公路或者快速路））、不属于州际公路系统的部分控制进入的所有城市主干道。

English	Chinese
Freight: See also Cargo, Commodity, Goods, Product.	**货物**：参见船货、日用品、商品、产品。
Freight: Property other than express and passenger baggage transported by air.	**货运**：除了特快和由飞机运送乘客行李之外的财物运送方式。
Freight: Any commodity being transported.	**货物**：被运送的任何商品。
Freight Agent: An establishment that arranges the transportation of freight and cargo for a fee. Revenue for freight agents (also known as shipping agents or brokers) represents commissions of fees and not the gross charges for transporting goods.	**货运经理处**：经营货物运输并收取费用的机构。货运经理处（也称为海运处或经纪人）所得收入是佣金的一种体现，而不是运输货物的总费用。
Freight All Kinds (FAK): Goods classified FAK are usually charged higher rates than those marked with a specific classification and are frequently in a container which includes various classes of cargo.	**不分货种货物（FAK）**：被划分为 FAK 的货物通常会被收取比归为具体类别的货物更高的费用，常常被放在包括杂货集装箱里。
Freight and Other Transportation Services Forwarding: Includes establishments that provide forwarding, packing, and other services incidental to transportation. Also included are horse-drawn cabs and carriages for-hire.	**货物运送和其它运输服务代理**：包括提供代理、包装和其它运输附属服务的机构。也包括出租马车和出租货运车。
Freight Container: A reusable container having a volume of 64 cubic feet or more, designed and constructed to permit being lifted with its contents intact and intended primarily for containment of packages (in unit form) during transportation.	**货物集装箱**：容积大于或等于 64 立方英尺、主要设计用来在运输中装运包裹（统一形状），并且可以在货物完整无缺的情况下被挪动的容器。

English	Chinese
Freight Forwarder: An individual or company that accepts less-than-truckload (TLT) or less-than-carload (LCL) shipments from shippers and combines them into carload or truckload lots. Designated as a common carrier under the Interstate Commerce Act.	**货运代理商**：从不同托运人接收不足一卡车（TLT）或者不足一汽车容量（LCL）的货物，然后将它们合并装车进行运输的个人或公司。用来指州际商业法中的承运商。
Freight Forwarding: Establishments primarily engaged in undertaking the transportation of goods from shippers to receivers for a charge covering the entire transportation, and in turn making use of the services of various freight carriers in effecting delivery. Establishment pays transportation charges as part of its costs of doing business and assumes responsibility for delivery of the goods. There are no direct relations between shippers and the various freight carriers performing the movement.	**货运代理公司**：主要从事发货人和收货人之间的货物运输并收取整个运输所涵盖的费用以及使用各种与货物运输有关的服务的费用的公司。作为商业运作费用的一部分，货运代理公司支付交通费用，并承担货物递送的义务。在托运人和各种承担货物运输的运输方之间没有直接的联系。
Fuel: See also Gasohol, Gasoline, Kerosene.	**燃料**：参见汽醇，汽油，煤油。
Fuel: The primary fuel or energy source delivered to a residential site. It may be converted to some other form of energy at the site. Electricity is included as a fuel. Other primary fuels are coal, fuel oil, kerosene, liquefied petroleum gas (LPG), natural gas, wood, and solar.	**燃料**：送往居民点的主要燃料或能源。在居民点它可以被转化成其它能源形式。电力包含在燃料之内。其它主要的燃料有煤、燃料油、煤油、液化石油气（LPG）、天然气、木头和太阳能。
Fuel Cell: A device that produces electrical energy directly from the controlled electrochemical oxidation of the fuel. It does	**燃料电池**：从燃料的受控电化氧化中直接制造电能的设备。与大多数其它发电技术不同，

not contain an intermediate heat cycle, as do most other electrical generation techniques.

它不包含中间热循环过程。

Fuel Code: A 2-digit numeric code that identifies the type of fuel used. The code identifies regular (gasoline and diesel) fuels, alternative fuels such as natural gas and methanol, and vehicles able to operate on a combination of these fuels (regular and alternative).

燃料代码：标识所使用的燃料类型的两位数字代码。代码标识普通燃料（汽油和柴油）、代用燃料（比如天然气和甲醇）以及能够混合使用这些燃料（普通燃料和代用燃料）的车辆。

Fuel Fire/Explosion: Accidental combustion of vessel fuel, liquids, including their vapors, or other substance such as wood or coal.

燃料起火/爆炸：船舶燃、液体，包括它们的蒸汽，或者其它物质如木料或煤的意外燃烧或爆炸。

Fuel Injection: (See also Carburetor, Diesel Fuel System) A fuel delivery system whereby gasoline is pumped to one or more fuel injectors under high pressure. The fuel injectors are valves that, at the appropriate times, open to allow fuel to be sprayed or atomized into a throttle bore or into the intake manifold ports. The fuel injectors are usually solenoid operated valves under the control of the vehicle's on-board computer (thus the term "electronic fuel injection"). The fuel efficiency of fuel injection systems is less temperature dependent than carburetor systems. Diesel engines always use injectors.

燃油喷射：（参见汽化器，柴油系统）汽油被高压注入一个或多个燃油注射器的燃油输送系统。燃油注射器设有阀门，在合适的时间开启让燃油喷射或雾射进入节流孔或者进气歧管端口。燃油注射器为受车辆的车载电脑控制的电磁阀（参见术语"电子燃油注射"）。燃油注射系统的燃油效率与汽化器系统相比受温度的影响较小。柴油机一直使用喷射器。

Fuel Tank: A tank other than a cargo tank, used to transport flammable or combustible liquid, or compressed gas for the purpose of supplying fuel for propulsion of the transport vehicle to which it is attached, or for the

油箱：与货舱不同，它是用来运输可燃或易燃液体或者压缩气体的容器。其目的是向油箱所依附的车辆本身的驱动设备或者车辆之上的其它设备的运

English	Chinese
operation of other equipment on the transport vehicle.	营补给燃油。
Fuel Tank Fitting: Any removable device affixed to an opening in the fuel tank with the exception of the filler cap.	**油箱接头**：除加油口盖外，附于油箱口部的可移动设备。
Fueling: Any stage of the fueling operation; primarily concerned with introduction of explosive or combustible vapors or liquids on board.	**加油**：加燃油操作中的任何阶段，主要指将易爆易燃气体或液体导入到飞机和车船上相关设备。
Full Trailer: Any motor vehicle other than a pole trailer which is designed to be drawn by another motor vehicle and so constructed that no part of its weight, except for the towing device, rests upon the self-propelled towing unit. A semitrailer equipped with an auxiliary front axle (converter dolly) shall be considered a full trailer.	**拖车**：除长货挂车之外、设计用来被另一个机动车辆拖拉的任何机动车辆，除牵引设备之外的全部重量都没有自驱动牵引装置。具有辅助前轴的半拖车应（牵引台车）视为拖车。
Full Trailer: A truck-trailer with front and rear axles. The load weight is distributed over both the front axle(s) and rear axle(s).	**双轴全挂车**：具有前后轴的货车挂车。载重分布于前轴和后轴。
Fullmount: A smaller vehicle mounted completely on the frame of either the first or last vehicle in a saddlemount combination.	**全固小车**：完全固定在的第一辆车或者最后一辆车的鞍形联接车架上的较小车辆。
Furniture Van Body: Truck body designed for the transportation of household goods; usually a van of drop-frame construction.	**家具敞蓬车车身**：设计用来运输家庭用品的卡车车身，通常是没有车架结构的敞蓬车。

English	Chinese
Gallon: (See also Barrel) A volumetric measure equal to 4 quarts (231 cubic inches) used to measure fuel oil. One barrel equals 42 gallons.	**加仑**：（参照桶）容积衡量。1 加仑 = 4 夸脱（231 立方英寸）。一般用于衡量液体和燃油，此外，一桶等于 42 加仑。
Gantry: A frame structure raised on side supports so as to span over or around something.	**千斤顶**：由旁侧支起并跨越中间的框架结构。
Gap: Low point or opening between hills or mountains or in a ridge or mountain range.	**裂口**：低点或两山陵或山脉之间的空隙。
Garage: A space large enough to accommodate a car, with a door opening at least six feet wide and seven feet high. "Attached" means it is under part or all of the house or it shares part of a wall in common with the house. Not included are carports, barns, or buildings (not connected to the house) or storage space for golf carts or motorcycles.	**车库**:大小够放置一辆汽车的空间，一个打开后至少 6 英尺宽，7 英尺高的门。附属型车库是指它在部分或全部房子的下面，或和房子共用部分墙。不包括无墙车库、棚舍、独立建筑，或者放置高尔夫球车和摩托车的储存间。
Garbage and Trash Collection: Establishments primarily engaged in collecting and transporting garbage, trash, and refuse, within a city, town, or other local area, including adjoining towns and suburban areas.	**垃圾和废品收集**：一个专门负责市区，郊区，或邻近城镇的垃圾收集和运输业务的机构。
Gas: (Except when designated as inert) Natural gas, other flammable gas, or gas which is toxic or corrosive.	**气体**：(除限定为惰性外)天然气等可燃气体，或是有毒或带腐蚀性的气体。
Gas: A non-solid, non-liquid combustible energy source that includes natural gas,	**燃气**:非固体非液体可燃性能源，包括天然气，焦炉煤气，

English	Chinese
coke-oven gas, blast-furnace gas, and refinery gas.	高炉煤气，炼油厂气。
Gas Guzzler Tax: Originates from the 1978 Energy Tax Act (Public Law 95418). A new car purchaser is required to pay the tax if the car purchased has a combined city/highway fuel economy rating that is below the standard for that year. For model years 1986 and later, the standard is 22.5 mpg.	**高耗油税**：起源于 1978 年能源税法（美国公共法案 95418）。凡新购买的轿车其城市或高速公路的综合燃油经济指标低于该年度标准的,购买者需付高耗油税。自 1986 年以来，综合燃油经济指标是 22.5 英里/加仑。
Gear Bonger: Driver who grinds gears when shifting.	**换档迟疑者**：在换档时经常让齿轮互相摩擦的司机。
Gear Jammer: One who constantly clashes the gears.	**换档过猛者**: 在换档时经常让齿轮卡住的司机。
Gear Ratio: The number of revolutions a driving gear requires to turn a driven gear one revolution. For a pair of gears, the ratio is found by dividing the number of teeth on the driven gear by the number of teeth on the driving gear.	**齿轮比**：将被动齿轮驱动一周所需的主动轮转动的转数。一对齿轮的齿轮比等于被动齿轮数除以主动齿轮数。
General Freight Carrier: A carrier which handles a wide variety of commodities.	**普通货机**：负责运载各种各样商品的承运方。
General Freight Carrier: Trucking company engaged in shipping packaged, boxed, and palletized goods that can be transported in standard, enclosed tractor-trailers, generally 40 to 48 feet in length.	**普通货物承运方**：从事承运包装、箱装和托盘装载的货运卡车公司，所承运的货物可以装入标准的 40-48 英尺长的封闭式卡车。
General Warehousing and Storage: Establishments primarily engaged in the	**通用仓库**：用于储存普通货物的仓库。

English	Chinese

warehousing and storage of a general line of goods.

Geographical Information System (GIS): A system of hardware, software, and data for collecting, storing, analyzing, and disseminating information about areas of the Earth. For Highway Performance Monitoring System (HPMS) purposes, Geographical Information System (GIS) is defined as a highway network (spatial data which graphically represents the geometry of the highways, an electronic map) and its geographically referenced component attributes (HPMS section data, bridge data, and other data including socioeconomic data) that are integrated through GIS technology to perform analyses. From this, GIS can display attributes and analyze results electronically in map form.

地理信息系统：一个集硬件、软件、和数据于一体的系统。用于采集、储存、分析、传播地理信息。例如，在公路监测系统（HPMS）中，地理信息系统被定义成高速公路系统（从地理上描述高速公路结构的空间数据，电子地图）和地理参数和特征（HPMS 地段，桥梁和社会经济数据）。他们通过地理信息系统技术进行整合，从而进行分析。可以以地图的方式显示出属性数据，并分析结果。

Glad Hands: Air hose brake system connections between tractor and trailer.

连接阀门：连接卡车头和拖车的空气软管刹车系统。

Global Positioning System (GPS): A space base radio positioning, navigation, and time transfer system being developed by the Department of Defense. When fully deployed, the system is intended to provide highly accurate position and velocity information, and precise time, on a continuous global basis, to an unlimited number of properly equipped users. The system will be unaffected by weather, and will provide a worldwide common grid

全球定位系统：由美国国防部开发的无线电定位导航和时间转化系统。当系统全部部署后，可以提供无间断的高精确度的位置和速度信息，以及精确时间信息。该服务无间断地提供给无数个有适当接收设备的用户。该系统不受天气影响，并提供全球通用的网络参照系统。GPS 的概念是基于准确和连续的信息，这一信息来

English	Chinese

reference system. The Global Positioning System (GPS) concept is predicated upon accurate and continuous knowledge of the spatial position of each satellite in the system with respect to time and distance from a transmitting satellite to the user. The GPS receiver automatically selects appropriate signals from the satellites in view and translates these into a three-dimensional position, velocity, and time. Predictable system accuracy for civil users is projected to be 100 meters horizontally. Performance standards and certification criteria have not yet been established.

就是每个卫星向用户传送的时间和空间位置信息。GPS 发射机自动地从视觉中的卫星选择适当的信号，并将其转化为三维空间的位置、速度、和时间信息。预计民用用户的精确度在水平面上为 100 米，性能标准和认证标准尚未制定。

Goat 'N' Shoat Man: Driver of a livestock carrier.

牧羊人：畜禽类货物运输者。

Good Condition Classification: No corrective maintenance is needed at time of inspection. Facility is serving the purpose for which it was constructed.

状况良好标准：在定期检查时，确认为不需要故障检修。设施能为建造目的服务。

Goods: See also Cargo, Commodity, Freight, Product.

货物：参照货物,日用品,货运和产品等。

Government Aid Cargo: The tonnes of cargo assessed at the Government aid rate of tolls as defined in the St. Lawrence Seaway Tariff of Tolls.

政府援助物资：根据圣劳伦斯运河（美国和加拿大）收费表评估出来的政府援助物资吨数。

Government Fleet Vehicle: Includes vehicles owned by all federal General Services Administration (GSA), state, county, city, and metro units of government, including toll road operations.

政府车队车辆：指（美国）联邦总务署、州政府、县、市、公交机构，包括高速公路收费处的运营车辆。

English	Chinese
Government Leased Vehicle: A vehicle obtained by an executive agency by contract or other source for a period of 60 continuous days or more.	**政府租用车辆**：由行政机构依照合同或其他方式连续租用六十天及以上的车辆
Government Owned Contractor Operated Vehicle: A vehicle that is owned or leased by the Federal Government but used by a contractor under a cost reimbursement contract with a Federal agency.	**政府拥有但由合同商使用的车辆**: 政府拥有或租用的车辆，由合同商使用，操作费用按照双方的合同来报销。
Government Owned Vehicle: A vehicle that is owned by the Federal Government.	**政府所拥有的车辆**：联邦政府拥有的车辆。
Government Transportation Expenditures: Expenditures are the final actual costs for capital goods and operating services covered by the government transportation program.	**政府运输支出**：政府运输计划所涵盖的运行操作及各项投资的最终实际支出。
Government Transportation Revenue: The transportation revenue estimates contained in this report consist of those funds identified as government transportation-related user charges, taxes or fees in the various data sources. Therefore, general revenue is not included.	**政府运输收入**：在交通统计局的报告中，对运输收入依据数个不同的数据来源进行估算，其构成包括养路费，税收及其他使用费。因此，普通年度收入不包括在此项中。
Governor: A device which limits the speed of an engine. A governor is also a part on an automatic transmission which signals internal transmission components to shift to a higher gear.	**调速器**：用于限制发动机速度的装置，同时也是汽车变速箱的一部分，用于内部传输信号让档位挂到更高一档。
Grab One: To shift into a lower gear as a means of gaining power when driving uphill.	**降档**：当车辆爬坡时，需要换到低档以增加爬坡的动能。

English	Chinese
Gradability: The ability of a vehicle to negotiate a given grade at a specified Gross Combination Weight Rating (GCWR) or Gross Vehicle Weight Rating (GVWR). It is the measure of the starting and grade climbing ability of a vehicle, and is expressed in percent grade, (1 percent is a rise of 1 foot in a horizontal distance of 100 feet).	**爬坡牵引力**：在给定的坡度下，具有一定的总组合重量级别或总车辆重量级别下车辆的爬坡能力。它用于衡量车辆的起步及爬坡能力，由百分比级别来表示（百分之一表示在 100 英尺的距离上升高一英尺）。
Grain Body: Low side, open top truck body designed to transport dry fluid commodities.	**谷类运载车厢**：一种运输干的流动物质的低边敞篷的货车车厢。
Grain Cargo: The tonnes of cargo assessed at the Food or Feed Grains rate of tolls as defined in the St. Lawrence Seaway Tariff of Tolls.	**谷类货物**：按美国和加拿大圣伦斯海运收费表中的规定，评估粮食或饲料的货物吨数
Grants: A federal financial assistance award making payment in cash or in kind for a specified purpose. The federal government is not expected to have substantial involvement with the state or local government or other recipient while the contemplated activity is being performed. The term "grants-in-aid" is commonly restricted to grants to states and local governments.	**资助**：一种用于特定目的的，以现金或其它方式支付的联邦政府财政资助。联邦政府不实质上参与州或地方政府或被资助单位的项目。 通常"资助"局限于对州或地方政府的资助。
Gross Axle Weight Rating (GAWR): The value specified by the vehicle manufacturer as the load carrying capacity of a single axle system, as measured at the tire-ground interfaces.	**总轴重级别**：（公路）由车辆制造商所制定的每轴载重量，在轮胎和地面的接触面上计量。

English	Chinese
Gross Combination Weight (GCW): The maximum allowable fully laden weight of a tractor and its trailer(s).	**总重量**：（公路）车头加最大载荷拖车的总重量。
Gross Combination Weight Rating (GCWR): The value specified by the manufacturer as the loaded weight of a combination (articulated) vehicle. In the absence of a value specified by the manufacturer, GCWR will be determined by adding the Gross Vehicle Weight Rating (GVWR) of the power unit and the total weight of the towed unit and any load thereon.	**总重量级别**：（公路）由车辆制造商所制定的组合车辆的载重级别。当制造商没有指定这一级别时，这一级别将由车辆动力装置的总重量级别加上拖车的总重量以及其他任何载重来决定。
Gross Domestic Product (GDP): The total value of goods and services produced by labor and property located in the United States. As long as the labor and property are located in the United States, the supplier (that is, the workers and, for property, the owners) may be either U.S. residents or residents of foreign countries.	**国内生产总值**：美国资产和劳力所生产出来的商品和服务的总价值。只要劳力和资产在美国境内，提供者（劳动者，资产的所有者）可以是美国居民，也可以是外国居民。
Gross Horsepower: The power of a basis engine at a specified revolution per mile (RPM) without alternator, water pumps, fan, etc. Gross horsepower is the figure commonly given as the horsepower rating of an engine.	**总马力**：除发动机、水泵、风扇等外，在一定转速下（每分钟转速 RPM），一个发动机所产生的功力。总马力是衡量发动机功率所广泛采用的标准。
Gross National Product (GNP): A measure of monetary value of the goods and services becoming available to the nation from economic activity. Total value at market	**国家生产总值**：经济活动中产生的可用于本国的商品和服务总货币值。指国家经济产生的所有的商品和服务的市场价

English	Chinese
prices of all goods and services produced by the nation's economy. Calculated quarterly by the Department of Commerce, the Gross National Product is the broadest available measure of the level of economic activity.	格。由商业部按季度计算，国家生产总值是衡量经济活动水平的最广泛应用的衡量指标。
Gross Vehicle Weight (GVW): The maximum allowable weight in pounds or tons that a truck is designed to carry.	**车辆总重**：（公路）卡车设计所允许的最大载重量。以磅或吨衡量。
Gross Vehicle Weight (GVW): The weight of the empty vehicle plus the maximum anticipated load weight.	**车辆总重**：（公路）车辆自重加上最大预期载荷。
Gross Vehicle Weight Rating (GVWR): The maximum loaded weight in pounds of a single vehicle. Vehicle manufacturers specify the maximum Gross Vehicle Weight Rating (GVWR) on the vehicle certification label.	**车辆总重级别**：一辆车载货时的总重量，车辆制造商会将这一最大载重级别标明在车辆的出厂标签上。
Gross Vehicle Weight Rating (GVWR): The maximum rated capacity of a vehicle, including the weight of the base vehicle, all added equipment, driver and passengers, and all cargo loaded into or on the vehicle. Actual weight may be less than or greater than GVWR.	**车辆总重级别**：（公路）车辆的总栽重能力，包括车重，所有车载设备，驾驶员，旅客，和所有货物的重量。实际重量有可能小于或大于车辆总重级别。
Gross Weight: (See also Net Weight) Entire weight of goods, packing, and container, ready for shipment.	**总重**：（参见净重）已经装箱的可发货的货物的总重量，包括货物本身，包装和集装箱的重量。
Gross Weight/Mass: The weight of a packaging plus the weight of its contents.	**总重/质量**：包装和货物的重量和。

English	Chinese
Ground: The flat horizontal surface on which the tires of a motor vehicle rest.	地面：车辆轮胎接触的地平面。
Ground Surface: The land surface of the earth, both exposed and underwater.	地表：地球表面，包括地面和水下表层。
Ground Visibility: Prevailing horizontal visibility near the earth's surface as reported by the United States National Weather Service or an accredited observer.	地面能见度：根据美国气象中心或其它权威机构所报道的接近地球表面的水平能见度。
Guard Rail: A strong fence or barrier to prevent vehicles from leaving the roadway, or for people's safety.	护拦：牢固的拦杆,用于防止车辆驶出道路以保护人生安全。
Gum Ball Machine: Rotating warning light on top of an emergency vehicle.	胶球机：安装在应急车辆顶部的旋转灯。
Gypsy: 1) An independent truck operator who drives his own truck and secures freight wherever he can. 2) One who trip leases to authorized carriers.	吉普赛：1）独立经营的驾驶自己的车辆承接货运业务的卡车司机。2）连卡车一起租借给指定承运人的卡车司机。

English	Chinese
H Point: The mechanically hinged hip point of a manikin which simulates the actual pivot center of the human torso and thigh, described in Society of Automotive Engineers (SAE) Recommended Practice J826, "Manikins for Use in Defining Vehicle Seating Accommodations," November 1962.	**H 点:** 机械铰链关节点。用来模拟人类躯干的枢纽系统,在《汽车工程学会实践》 J826 上"假人在确定车辆座椅座位上的应用"一文中有详尽描述。1962 年 11 月。
Hand Operated Switch: A non-interlocked switch which can only be operated manually.	**手控开关**:非互锁的开关。只能用手动操作。
Haulage Cost: Cost of loading ore at a mine site and transporting it to a processing plant.	**运输成本**:矿场装卸矿石以及运输矿石到加工厂的费用。
Hauling Post Holes: Driving an empty truck or trailer.	**空载**:开空卡车或拖车。
Hazard Warning Signal: Lamps that flash simultaneously to the front and rear, on both the right and left sides of a commercial motor vehicle, to indicate to an approaching driver the presence of a vehicular hazard.	**危险警告信号灯**:安装在车辆前后两侧的橙色警示灯。当打开此警示灯时警示灯可同步瞬间闪动以达到提示来往车辆前方有抛锚车辆障碍物的效果。
Hazard Zone: One of four levels of hazard (Hazard Zones A through D) assigned to gases, as specified in 49 CFR 173.116(a), and one of two levels of hazards (Hazard Zones A and B) assigned to liquids that are poisonous by inhalation, as specified in 49 CFR 173.133(a) of this subchapter. A hazard zone is based on the LC50 value for acute inhalation toxicity of gases and vapors, as specified in 49 CFR 173.133(a).	**有害危险区域**: 四个检测有害危险气体程度的等级之一(从 A 到 D),见章节 49 CFR 173.116(a)。该定义也是两个检测经由吸入的有毒液体等级的标准之一(A 和 B),见 49 CFR 173.133(a)。 有害危险区域的界定是基于急性吸入有毒的气体和蒸汽的 LC50 值,见 49 CFR 173.133(a). 。
Hazardous Material (HAZMAT): (See also Highly Volatile Liquid, Marine Pollutant) A	**有害危险物质 (HAZMAT):** (另见"高挥发性液体"和"海洋

English	Chinese
substance or material which has been determined by the Secretary of Transportation to be capable of posing an unreasonable risk to health, safety, and property when transported in commerce, and which has been so designated. The term includes hazardous substances, hazardous wastes, marine pollutants, and elevated temperature materials as defined in this section, materials designated as hazardous under the provisions of 49 CFR 172.101 and 172.102, and materials that meet the defining criteria for hazard classes and divisions in 49 CFR 173.	水污染物"）。按照美国联邦交通部的规范，凡在以商业形式运输流通的过程中具备并可能对健康，安全及财产构成一定危害的物质或材料都属于有害危险物质。该界定范围包括在 49 CFR 172.101 and 172.102，49 CFR 173. (49CFR171) (49CFR390) 等附属章节中所定义的所有有害物质，有害废物，海洋污染物和高温易蒸发性材料。
Hazardous Material (HAZMAT) Employee: A person who is employed by a HAZMAT employer and who in the course of employment directly affects hazardous materials transportation safety. This term includes an owner-operator of a motor vehicle which transports hazardous materials in commerce. This term includes an individual, including a self-employed individual, employed by a HAZMAT employer who, during the course of employment: 1) Loads, unloads, or handles hazardous materials; 2) Tests, reconditions, repairs, modifies, marks, or otherwise represents containers, drums, or packagings as qualified for use in the transportation of hazardous materials; 3) Prepares hazardous materials for transportation; 4) Is responsible for safety of transporting hazardous materials; or	**有害危险物工作者**：受雇于经营有害危险物商品（HAZMAT）的厂家并且在雇佣期间接触并能直接影响危险物运输安全的工作者。该界定包括驾驶运输车辆的受雇员工以及雇主本人。受雇员工（包括个体经营者）在受雇期间运输有害危险物品的活动包括：1）装卸载或处理危险物品，2）检测，调试及修正有害危险物品技术参数和指标；在容器和包装桶上粘贴与包装运输过程所要求的各种规范的标识物及标签，3）办理运输危险物品，4）运输危险品的安全负责或，5）驾驶用于运输危险品的车辆。

English	Chinese

5) Operates a vehicle used to transport hazardous materials.

Hazardous Material (HAZMAT) Employer: A person who uses one or more of its employees in connection with: transporting hazardous materials in commerce; causing hazardous materials to be transported or shipped in commerce; or representing, marking, certifying, selling, offering, reconditioning, testing, repairing, or modifying containers, drums, or packagings as qualified for use in the transportation of hazardous materials. This term includes an owner-operator of a motor vehicle which transports hazardous materials in commerce. This term also includes any department, agency, or instrumentality of the United States, a State, a political subdivision of a State, or an Indian tribe engaged in an activity described in the first sentence of this definition.

有害危险物雇主：雇用一个或多个员工从事有害危险物商品运输的商业活动业主。其商业活动的内容包括运送有害危险物品；检测，调试及认正有害危险物品技术参数和指标；销售或提供有害危险物；以及对运送有害危险物的容器和包装筒的检测，调试，维修维护及改装，在容器和包装桶上粘贴与包装运输过程所要求的各种规范的标识物及标签。该规范的界定范围也包括经营有害危险物品运输的个体经营者。该规范的界定适用于联邦政府管辖的所有部门，机构，与其有业务往来的单位，各州的相关部门，各州的政治团体，以及在印第安部落保护区内从事符合本规范所定义的的活动。

Hazardous Material Residue: The hazardous material remaining in a packaging, including a tank car, after its contents have been unloaded to the maximum extent practicable and before the packaging is either refilled or cleaned of hazardous material and purged to remove any hazardous vapors.

有害物质残留物：在最大程度清空有害危险物之后，**残留在**容器，包装筒，或运输罐车内的有害危险物质。

Hazardous Materials: Any toxic substance, explosive, corrosive material, combustible material, poison, or radioactive material that

危险材料：任何有害物质，诸如爆炸性，腐蚀性，易燃物质，毒药或放射性材料，在商

English	Chinese

poses a risk to the public's health, safety, or property when transported in commerce.

业运输过程中构成对公众健康，安全，或财产的威胁。

Hazardous Substance: A material, and its mixtures or solutions, that 1) Is identified in the appendix to 49 CFR 172.101; 2) Is in a quantity, in one package, which equals or exceeds the reportable quantity (RQ) listed in the Appendix A to 49 CFR 172.101; and 3) When in a mixture or solution which, for radionuclides, conforms to paragraph 6 of Appendix A, or, for other than radionuclides, is in a concentration by weight which equals or exceeds the concentration corresponding to the RQ of the material, as shown in the Table appearing in 49 CFR 171.8. This definition does not apply to petroleum products that are lubricants or fuels.

有害物质：一种物质，不管是混合物或溶液，即 1）在附录 49 CFR 172.101 中确定; 2）一个数量包等于或超过该报告值（RQ）在附录 A 至 49 CFR 172.101 列出; 3）当在混合物或溶液，放射性核素，遵照段落 6 附录 A，非放射元素，根据重力集结，等于或超过相对应的 RQ 材料的浓度，如表 49 CFR 171.8 所示。这个定义不适用于石油产品如润滑剂和燃料。

Hazardous Waste: Any material that is subject to the hazardous waste manifest requirements of the Environmental Protection Agency (EPA) specified in 40 CFR 262 or would be subject to these requirements absent an interim authorization to a State under 40 CFR 123, subpart F.

危险废物：任何符合环境保护局（EPA）在 40 CFR 262 清单中对危险废物指定的要求，或符合 40 CFR 123, subpart F.申明中的要求而没有州属临时许可证危险物品。

Head Impact Area: All nonglazed surfaces of the interior of a vehicle that are statically contactable by a 6.5-inch diameter spherical head form of a measuring device having a pivot point to "top-of-head" dimension infinitely adjustable from 29 to 33 inches in accordance with the procedure explained in 49 CFR 390.

头部冲击地区：所有没有上釉的车的内部的表面，从统计角度可被测量的 6.5 英寸直径的球形头部接触区域。49 CFR 390.5 中程序的解释，此区域从支点到顶层，从 29 至 33 英寸可以进行无限量的调整。

English	Chinese
Head Lamps: Lamps used to provide general illumination ahead of a motor vehicle.	头灯：机动车前面用于提供一般照明的灯。
Head On Collision: 1) Refers to a collision where the front end of one vehicle collides with the front-end of another vehicle while the two vehicles are traveling in opposite directions. 2) A collision in which the trains or locomotives involved are traveling in opposite directions on the same track.	车头碰撞：1）碰撞指相反方向行驶的两辆车一个车的前端与另一辆车的前端相撞。2）也可以是列车或机车在相反方向在同一轨道上相撞。
Headache Rack: Heavy bulkhead that extends over cab from trailers, usually made of pipe and used in steel hauling.	运架：重型分壁拖车一直延伸到机车室，通常由管道制造并用于钢材运输。
Header Bar: Rear cross piece on open top trailer.	横梁：开放的拖车顶部后面横跨车身的金属杠。
Header Board: Protective shield at front end of flat bottom trailer to prevent freight from shifting forward.	横梁板：平底拖车前端保护罩用来防止货物向前移动。
Heater: Any device or assembly of devices or appliances used to heat the interior of any motor vehicle. This includes a catalytic heater which must meet the requirements of 49 CFR 177.834(1) when flammable liquid or gas is transported.	电热器：任何设备或成套的设备或器具用于加热汽车内部。当运输易燃液体或气体时，这包括必须满足 49 CFR 177.834 条款的催化热水器（1）。
Heavy Duty Truck: Truck with a gross vehicle weight generally in excess of 19,500 pounds (class 6-8). Other minimum weights are used by various law or government agencies.	重型卡车：卡车的总重超过 19,500 英镑（6-8 级）以上。其他最低重量由不同的法律或政府机构使用。

English	Chinese
Heavy Hauler Trailer: A trailer with one or more of the following characteristics: 1) Its brake lines are designed to adapt to separation or extension of the vehicle frame; or 2) Its body consists only of a platform whose primary cargo carrying surface is not more than 40 inches above the ground in an unloaded condition, except that it may include sides that are designed to be easily removable and a permanent "front-end structure" as that term is used in 49 CFR 393.106.	**重型运输机拖车**：拖车有以下一个或多个特点：1）它的制动管路的设计用来适应分开或车架延伸；2）车体只包括一个平台，其主要的货物运输面，在卸载的情况下，距离地面不超过 40 英寸高，除非它包括设计用于方便地移动的边和一个永久性的"前端结构"。这一用语用于 49 CFR 393.106。
High-Mileage Households: Households with estimated aggregate annual vehicle mileage that exceeds 12,500 miles.	**高里程用户**：大概每年的机动车使用总里程超过一万二千五百英里的家庭。
High Speed Rail: A rail service having the characteristics of intercity rail service which operates primarily on a dedicated guideway or track not used, for the most part, by freight, including, but not limited to, trains on welded rail, magnetically levitated (MAGLEV) vehicles on a special guideway, or other advanced technology vehicles, designed to travel at speeds in excess of those possible on other types of railroads.	**高速铁路**：铁路系统服务城际铁路经营主要在专用导轨，大部分不用来货运，包括但不限于焊轨列车，特定轨道上的磁悬浮其他先进技术的车辆，设计用于以超出其它类型铁路的速度运行。
High Speed Rail: A rail transportation system with exclusive right-of-way which serves densely traveled corridors at speeds of 124 miles per hour and greater.	**高速铁路**：铁路系统中有专用的轨道用来服务高密度的线路。速度一般在每小时 124 英里或更高。
High Type Road Surface: High flexible, composite, rigid, etc. (Surface/Pavement Type Codes 61, 62, 71-76 and 80).	**高标准路面**：高弹性，复合材料，刚性等（表面/路面类型代码 61, 62, 71-76 and 80）。

English	Chinese
High Volume Area: An area which an oil pipeline having a nominal outside diameter of 20 inches or more crosses a major river or other navigable waters, which, because of the velocity of the river flow and vessel traffic on the river, would require a more rapid response in case of a worst case discharge or substantial threat of such a discharge.	**高容量面积**：一个地区石油管道具有大于等于 20 英寸的外径，穿过主要河流或其他通航水域，由于河水流量和内河船舶交通速度，将需要更多的快速反应，在遇到如排放或泄漏威胁的情况下。
Highly Volatile Liquid: (See also Hazardous Material) A hazardous liquid which will form a vapor cloud when released to the atmosphere and which has a vapor pressure exceeding 276 kPa (40 psia) at 37.8 C (100 F).	**高挥发性液体**：（另见有害物质）一种有害液体将形成蒸气云释放到大气中，并且蒸气压超过 276 千帕 (40 平方英寸) 在 37.8 C (100 F)。
Highway: Is any road, street, parkway, or freeway/expressway that includes rights-of-way, bridges, railroad-highway crossings, tunnels, drainage structures, signs, guardrail, and protective structures in connection with highways. The highway further includes that portion of any interstate or international bridge or tunnel and the approaches thereto (23 U.S.C. 101a).	**高速公路**：任何道路，街道，公路，或高速公路/快速公路，有优先行驶权，桥梁，铁路和高速路交叉，公路隧道，隧道，排水结构，标志，护栏，和保护物用来连接高速路。高速路还包括，任何州际或国际桥梁或隧道和接入段 (23 U.S.C. 101a)。
Highway Construction Project: A project financed in whole or in part with Federal-aid or Federal funds for the construction, reconstruction or improvement of a highway or portions thereof, including bridges and tunnels.	**公路建设项目**：全部或部分由联邦资助的一个项目，改建或改进公路中的某些部分，包括桥梁和隧道。
Highway Mode: Consists of public roads	**高速公路模式**：包括公共道路

English	Chinese

and streets, automobiles, vans, trucks, motorcycles, and buses (except local transit buses) operated by transportation companies, other businesses, governments, and households, garages, truck terminals, and other facilities for motor vehicles.

和街道，汽车，面包车，卡车，摩托车，运输公司运营的公共汽车（除当地的公共巴士），其他企业，政府和家庭经营，车库，卡车中转站，以及其他机动车辆设施。

Highway Performance Monitoring System (HPMS): The State/Federal system used by the FHWA to provide information on the extent and physical condition of the nation's highway system, its use, performance, and needs. The system includes an inventory of the nation's highways including traffic volumes.

公路性能监控系统（HPMS）：联邦公路管理局使用的国家/联邦系统，用来提供的全国公路系统程度和物理状态的信息，它的使用，性能和需要。该系统包括一个全国的交通量清单。

Highway-Rail Crossing: (See also At Grade, Grade Crossings) A location where one or more railroad tracks intersect a public or private thoroughfare, a sidewalk, or a pathway.

公路与铁路交叉：（又见级，平面交叉）一个或多个铁路轨道与公共或私人通道，人行道或线路相交的位置。

Highway-Rail Crossing Accident/Incident: An impact between on track railroad equipment and a highway user (e.g., an automobile, bus, truck, motorcycle, bicycle, farm vehicle, pedestrian or other highway user) at a designated crossing site. Sidewalks, pathways, shoulders and ditches associated with the crossing are considered to be part of the crossing site. The term "highway user" includes pedestrians, cyclists, and all other modes of surface transportation.

公路与铁路交叉事故/事件：铁路轨道上的设备和高速公路用户（例如，汽车，公共汽车，卡车，摩托车，自行车，农用车，行人或其它公路使用者）在指点的交叉口互相发生作用。人行道，通道，路肩和与沟渠和交叉口相关，也看成是交叉口站点的一部分。术语"公路使用者"，包括行人，骑自行车和使用地面其它交通工具的人。

English	Chinese
Highway Research Information Service (HRIS): A computer-based information storage and retrieval system developed by the Transportation Research Board with financial support from the state highway and transportation departments and the Federal Highway Administration. It consists of summaries of research projects in progress and abstracts of published works.	**公路研究信息服务（HRIS）：** 一种基于计算机的信息存储和检索系统，由运输研究委员会开发的，得到国家公路和运输部门以及联邦公路管理局的财政支持。它包括目前进行中的项目进展和出版作品的摘要。
Highway Trust Fund: The federal account established by law to hold receipts collected by the government and earmarked for highway programs and a portion of the federal mass transit program. It is supported by the federal gasoline tax and other user taxes.	**高速公路信托基金：根据**法律制定的联邦帐户，用于持有政府收取的收入，是专用公路项目和政府公共运输计划的一部分。它得到了联邦汽油税和其他用税的支持。
Highway User Fee or Tax: A charge levied on persons or organizations based on the use of public roads. Funds collected are usually applied toward highway construction, reconstruction and maintenance. Examples include vehicle registration fees, fuel taxes, and weight-distance taxes.	**公路使用费或税：** 基于对公共道路的使用，由收费人员或组织收取的费用。募集的资金通常用于对公路建设，重建和维修。例子包括：车辆登记费，燃油税，重量距离税。
Hit and Run: (See also Crash) A hit-and-run occurs when a motor vehicle in transport, or its driver, departs from the scene after being involved in a crash prior to police arrival on the scene. Fleeing pedestrians and motor vehicles not in transport are excluded from the definition. It does not matter whether the hit-and-run vehicle was striking or struck.	**交通肇事后逃逸：** （亦见撞车事故）交通肇事后逃逸，当机动车或机动车司机，在警方到达现场之前从现场离开。交通行人和不在运行中的车辆不在定义内。不论逃逸车辆是撞车的还是被撞得。

English	Chinese
Hobo: Tractor that is shifted from terminal to terminal.	HOBO：没有分配给固定终端的牵引机。
Holding Agency: A federal agency having accountability for motor vehicles owned by the Government. This term applies when a federal agency has authority to take possession of, assign or reassign motor vehicles regardless of which agency is using the motor vehicles.	控股机构：一个对政府拥有机动车量负责的联邦机构，。该术语适用于当一个联邦机构有权接管，分配或重新分配汽车无论哪个机构使用机动车辆。
Home Base: The location where a vehicle is usually parked when not in use or on the road.	停车位置：一般凡车辆在不使用时停车的地点。
Home Signal: A roadway signal at the entrance to a route or block to govern trains in entering and using that route or block.	进站信号：位于入口处的公路路线信号或阻止列车进入和使用该线路。
Hood Lifter: Garage mechanic.	车罩升降器：车库工具。
Hopper: A top loading, funnel-shaped structure for temporary storage of loose materials, which will be dispensed from the bottom.	漏斗：顶部装载，漏斗型的结构，用于临时储存松散材料一般从底部配发。
Hopper Body: Truck body capable of discharging its load through a bottom opening without tilting.	漏斗机构：卡车车体能释放力通过底部倾斜。
Horse: Tractor or power unit.	马力：牵引机或功的单元。
Horse Light: Spotlight mounted on cab to reveal open-range livestock.	马灯：安装在车头上的聚光灯，用以看到视线内的动物
Horse Van Body: Truck designed for the transportation of valuable horses (livestock).	运送马匹的厢式汽车：为名贵马匹设计的运输卡车。

English	Chinese
Horsepower (HP): The amount of work that an engine can perform within a given time. One horsepower equals 33,000 foot pounds of work per minute.	**马力（HP）**：在一定时间内由一个引擎牵引能做的总功。一马力等于每分钟 3.3 万英尺磅功。
Hot-Deck Imputation: A statistical procedure for deriving a probable response to a questionnaire item concerning a household or vehicle, where no response was given during the survey. To perform the procedure, the households or vehicles are sorted by variables related to the missing item. Thus, a series of "sort categories" are formed, which are internally homogeneous with respect to the sort variables. Within each category, households or vehicles for which the questionnaire item is not missing are randomly selected to serve as "donors" to supply values for the missing item of "recipient" households or vehicles.	**估算法**：一种统计程序,用于推算涉及一个家庭或车辆的问卷中,在调查中没回答的问题的答复。为了进行这个程序，要先将与空缺答复有关的家庭或车辆按应变量排序。然后，一列"分类类属"组成后，排序的应变量在本组内相同。在同一组中，问卷调查中的没有缺失数据的家庭或车辆可用作"捐助者"给这些有缺失数据的家庭或车辆提供依据。
Hot Load: Rush shipment of cargo.	**快件**：**快速运载**货物输。
Household: A group of persons whose usual place of residence is a specific housing unit; these persons may or may not be related to each other. The total of all U.S. households represents the total civilian non-institutionalized population. Does not include group quarters (i.e., 10 or more persons living together, none of whom are related).	**家庭**：一些人的惯常居所是一组特定的家庭单位，这些人可能相关也可能没有血缘关系。所有家庭人口代表除了受管制人口外所有的人口。不包括集体宿舍（即 10 或更多的人生活在一起，都不相联系）。
Household Trip: One or more household members traveling together.	**家庭出行**：一个或多个家庭成员一同旅行。

English

Household Vehicle: A motorized vehicle that is owned, leased, rented or company owned and available to be used regularly by household members during the travel period. Includes vehicles used solely for business purposes or business owned vehicles if kept at home and used for the home to work trip, (e.g., taxicabs, police cars, etc.) which may be owned by, or assigned to, household members for their regular use. Includes all vehicles that were owned or available for use by members of the household during the travel period even though a vehicle may have been sold before the interview. cludes vehicles that were not working and not expected to be working within 60 days, and vehicles that were purchased or received after the designated travel day.

Housing Unit: A house, apartment, a group of rooms, or a single room occupied or intended for occupancy as separate living quarters. Separate living quarters are those in which the occupants do not live and eat with any other persons in the structure and which have either 1) direct access from the outside of the building or through a common hallway intended to be used by the occupants of another unit or by the general public, or 2) complete kitchen facilities for the exclusive use of the occupants. The occupants may be a single family, one person living alone, two or more families living together, or any other group of related or

Chinese

家用车型：机动车辆被拥有，租用，租赁或公司拥有可供现有的家庭成员经常在旅行期间使用。其中包括专为商业目的或企业拥有的车辆，如果他们存放在家里而且用于家庭到工作的出行。，（例如，出租车，警车等）可能为家庭成员拥有或经常性使用。包括所有那些在旅行期间为家庭所有或可由家庭成员可使用的车辆，即使车辆可能已被出售。不包括坏的车辆或 60 天内不能修好的车辆，以及在指定的旅行后购买或拥有的车辆。

房屋单位：住宅，公寓，房间组或单个占用的房间，或作为单独的宿舍。独立的生活区是指无论是居住还是吃饭，不与其它人共享：1）从建筑外走廊直接进入或者有一般的大厅被居住者公共使用，或 2）专用的完整的厨房设施。居民可能是一个家庭，一个人独居，两个或两个以上的家庭生活在一起，或任何有关或无关的居住在一起的人。

English	Chinese

unrelated persons who share living arrangements.

Hydraulic Head: The distance between the respective elevations of the upstream water surface (headwater) above and the downstream surface water (tailwater) below a hydroelectric power plant.

液压头：中上游水面（河源）段和下游地表水（尾水低于水力发电厂）之间的高度差。

Hydrocarbon: A compound that contains only hydrogen and carbon. The simplest and lightest forms of hydrocarbon are gaseous. With greater molecular weights, they are liquid, while the heaviest are solids.

碳氢化合物：化合物仅包含氢和碳。最简单和最轻的碳氢化合物形式就是气态的。随着分子量越大，它们变成液体，而最重的是固体。

English	Chinese

Identification: The official legends "For Official Use Only" and "US Government," and other legends showing either the full name of the department, establishment, corporation, or agency by which it is used, if such title readily identifies the department, establishment, corporation, or agency concerned.

识别：官方标志"公务使用"和"美国政府"，和其它标志指明其隶属某具体的部门，公司或机构，如果这样的标志能用来识别该部门，机制，公司或相关的机构。

Identification Lamps: Lamps used to identify certain types of commercial motor vehicles.

识别灯：用于识别商业机动车类型的灯。

Idle Thrust: The jet thrust obtained with the engine power control level set at the stop for the least thrust position at which it can be placed.

空转推力: 机车停滞时设置的最小推动力

Imports: Receipts of goods into the 50 States and the District of Columbia from foreign countries and from Puerto Rico, the Virgin Islands, and other U.S. possessions and territories.

进口：从国外，波多黎各，维尔京群岛，和其他美国领属地进入到 50 个州和哥伦比亚特区的货物。

Impounding Space: A volume of space formed by dikes and floors which is designed to confine a spill of hazardous liquid.

蓄水空间：由堤防和地层形成的为了限制有害液体泄漏形成的空间。

Impounding System: Includes an impounding space, including dikes and floors for conducting the flow of spilled hazardous liquids to an impounding space.

蓄水系统：包括一个蓄水空间，及其引导危险液体到蓄水空间的堤防和地层。

Improper Loading: Loading, including weight shifting, of a vessel causing

不当装载：装载中包括重量转移，造成不稳定的，有限的可

English	Chinese

English

instability, limited maneuverability, or dangerously reduced freeboard.

Improper Lookout: No proper watch; the failure of the operator to perceive danger because no one was serving as lookout, or the person so serving failed in that regard.

In Bulk: The transportation, as cargo, of property, except Class A and B explosives and poison gases, in containment systems with capacities in excess of 3,500 water gallons.

In-Use Mile Per Gallon (MPG): A Miles Per Gallon (MPG) that was adjusted for seasonality and annual miles traveled.

Incident: See also, Accident, Casualty, Collision, Crash, Derailment, Event, Fatality, Injury, Truck Crash.

Incline Railway: Railway used to traverse steep slopes.

Income Taxes For the Period: Provisions for Federal, State, local and foreign taxes, which are based upon net income.

Indicated Airspeed: The speed of an aircraft as shown on its pitot static airspeed indicator calibrated to reflect standard atmosphere adiabatic compressible flow at sea level uncorrected for airspeed system errors.

Chinese

操作性，或危险的降低了超高。

不当瞭望：没有妥善观察;操作者没察觉到危险因为没有人担任把风，或服务的人员没能尽责。

成批的：货物运输（除 A 级和 B 级的炸药和毒气) 在容量超过 3500 加仑的集中箱中运输。

实际的每加仑可行驶的英里数 (MPG): 经过季节性和年度里程数调整的每加仑行驶的英里数。

事件：另见意外，严重事故，碰撞，碰撞，出轨，人员死亡，受伤，货车碰撞。

倾斜铁路：穿越陡峭山坡的铁路。

期间所得税：以净收入为基础的供给联邦，州，本地和外国税。

表速：显示在飞机校准空速指示器的速度，经校准反映海平面标准大气压绝热气流，未对空速系统的错误进行修正。

English	Chinese
Indication: The information conveyed by the aspect of a signal.	信号 ：由信号灯表达的信息。
Indication Locking: [with respect to rail operations] Electric locking which prevents manipulation of levers that would result in an unsafe condition for a train movement if a signal, switch, or other operative unit fails to make a movement corresponding to that of its controlling lever, or which directly prevents the operation of a signal, switch, or other operative unit, in case another unit which should operate first fails to make the required movement.	**信号锁闭**：[关于铁路运输业务] 如果一个信号，开关或者其他操作装置没有遵从对应的控制作出响应，指示锁闭实行电动锁定，从而制止导致不安全列车运行情况的杠杆操作，或者直接制止信号，开关或者其他操作装置的运行，以防其他应该先行操作的装置无法完成所需的动作。
Inductor: A track element consisting of a mass of iron, with or without a winding, that simulates the train control, train stop, or cab signal mechanisms on the rail vehicle.	**感应器**：由铁为主（有无线圈均可）组成的轨道，用来模拟列车控制，列车停止，或轨道车辆上的机车信号机制。
Industrial Sector: Construction, manufacturing, agricultural and mining establishments.	**工业产业**：建筑，制造业，农业和采矿业的机构。
Industrial Track: A switching track serving industries, such as mines, mills smelters, and factories.	**工业轨道**：为工业，如矿山，冶炼厂和工厂服务的调车轨道线路路，。
Industry Track: A switching track, or series of tracks, serving the needs of a commercial industry other than a railroad.	**工业轨道**：一个转接轨道，或一系列的轨道，服务于商业行业而不同于一般铁路的需求。
Informal Factory Visit: A visit by Coast Guard personnel to a manufacturing facility to acquaint the manufacturer with the existence of the law, regulations, general	**非正式工厂参观**：海岸警卫人员参观工厂，让其了解现有法律，法规，和一般行政上的要求以及并对可能存在的违法进

English	Chinese

administrative requirements affecting him, and possible penalties for violations.

行惩罚。

Infrastructure: 1) In transit systems, all the fixed components of the transit system, such as rights-of-way, tracks, signal equipment, stations, park-and-ride lots, but stops, maintenance facilities. 2) In transportation planning, all the relevant elements of the environment in which a transportation system operates.

24．**基础设施**：1）在运输系统中，所有的公交系统的固定部分如专属道路、轨道、信号设备、车站、转乘站、停靠站和维修设施。2）在交通规划中，包括交通运输系统运行的所有相关的外部环境因素。

Initial Impact Point: The first impact point that produced property damage or personal injury, regardless of First or Most Harmful Event.

最初影响点：最初造成财产损失或人员伤害的接触点，不论第一或最严重的事件。

Initial Terminal: The starting point of a locomotive for a trip.

初始终端：一个机车的出发点。

Injury: 1) Bodily injury resulting from a motor vehicle accident. To qualify as an "injury," the injured person must require and receive medical treatment away from the accident scene. 2) Harm to a person resulting from a single event, activity, occurrence, or exposure of short duration.

损伤：1）车辆事故造成的身体伤害。要符合作为一个"伤害"，受伤的人必须离开事故现场接受治疗。2）由一个单一的事件，活动，情况或短期暴露造成的对人的危害。

Injury Accident: An accident for which at least one injury, but no fatalities, was reported.

伤害事故：至少一人受伤，但没有人员死亡的事故。

Injury Crash: A police-reported crash that involves a motor vehicle in transport on a trafficway in which no one died but at least one person was reported to have: 1) An

碰撞伤害：有警方报告的机动车碰撞事故，没有人死亡，但至少有一人据报道有：1）致残伤害; 2）可见但无致残伤害;

English	Chinese

incapacitating injury; 2) A visible but not incapacitating injury; 3) A possible, not visible injury; or 4) An injury of unknown severity.

3）可能但不可见的伤害；4）一个未知严重程度受伤。

Injury Index: Refers to the highest degree of personal injury sustained as a result of the accident.

损伤指数：指由事故造成对人伤害的程度最高的等级。

Injury Rate: The average number of nonfatal injuries per accident or per one hundred accidents.

损伤率：每起或每 100 起非致命事故的平均受伤人数。

Injury Severity: The police-reported injury severity of the occupant, pedestrian, or pedalcyclist (e.g., severe or fatal; killed or incapacitating; minor or moderate; evident, but not incapacitating; complaint of injury; injured, severity unknown; no injury).

损伤程度：警方报告的乘客，行人，或自行车行人受伤的严重程度（例如，严重或致命的;死亡或致残;轻微或中度，明显的，但不是失去能力，抱怨受伤，严重程度未知的，没有伤害）。

Inland Area: The area shoreward of the boundary lines defined in 46 CFR 7, except that in the Gulf of Mexico, it means the area shoreward of the lines of demarcation (COLREG lines) defined in 33 CFR 80.740-80.850. The inland area does not include the Great Lakes.

内陆地区：在 46 CFR 7 定义的边界线向岸地区，除了墨西哥湾，它指的是在 33 CFR 740-80.850。定义的划定界限（COLREG 线）.内陆地区不包括五大湖。

Inland Zone: The environment inland of the coastal zone excluding the Great Lakes, Lake Champlain, and specified ports and harbors on inland rivers. The term inland zone delineates an area of federal responsibilities for response actions. Precise boundaries are determined by agreements

内陆区：在内陆沿海区不包括大湖区，尚普兰湖，指定港口，港口对内陆河流。内陆区界定了联邦应急反应行动职责范围的地区。精确界线由环境保护局（EPA）和美国海岸防卫队（美国海岸警卫队）协议

English	Chinese
between the Environmental Protection Agency (EPA) and the United States Coast Guard (USCG) and are identified in Federal Regional Contingency Plans.	规定，并在联邦地区应急计划中指明。
Inner Packaging: A packaging for which an outer packaging is required for transport. It does not include the inner receptacle of a composite packaging.	**内包装**：一种需要外部包装以满足运输要求。它不包括复合包装的内部贮藏器。
Inner Receptacle: A receptacle which requires an outer packaging in order to perform its containment function. The inner receptacle may be an inner packaging of a combination packaging or the inner receptacle of a composite packaging.	**内贮器**：需要外包装的内贮容器，以履行其包装作用。内部容器可能是组合包装或复合包装。
Inspection and Weighing Services For Motor Vehicle Transport Facility: Establishments primarily engaged in the operation of fixed facilities for motor vehicle transportation, such as toll roads, highway bridges, and other fixed facilities, except terminals.	**对机动车交通检查和称重的服务设施**：主要对机动车运输运作的固定设施，如收费公路，公路桥梁和其他固定设施，机场除外。
Instrument: A device using an internal mechanism to show visually or aurally the attitude, altitude, or operation of an aircraft or aircraft part. It includes electronic devices for automatically controlling an aircraft in flight.	**仪器**：一种使用内部机制来显示视觉或听觉的方位，海拔，或航行部分操作的设备。它包括自动飞行控制飞机的电子设备。
Insulated Body: Truck or trailer designed for transportation of commodities at controlled temperatures. It may be equipped for refrigeration or heating.	**绝缘体**：带有温度控制设计的用来运输货物的卡车或拖车。可能配有制冷或加热装备。

English	Chinese
Insulated Rail Joint: A joint in which electrical insulation is provided between adjoining rails.	**绝缘铁轨接头**：在相邻轨道之间使电气绝缘的接头。
Insured and Principal: The motor carrier named in the policy of insurance, surety bond, endorsement, or notice of cancellation, and also the fiduciary of such motor carrier.	**投保人及受益人**：在保险协议，担保债券，授权书或取消通告书中，也是该乘运人信托书中指定的乘运人。
Interagency Fleet Management System (IFMS): The organizational title assigned to the General Services Administrations (GSA) interagency fleet operation which encompasses the Central Office, Regional Offices, and all Fleet Management Centers and Fleet Management Subcenters.	**跨部门车辆管理系统**（IFMS）：分配给总事务行政局（GSA）的跨部门车辆运营的组织职权，包括中央办公厅，地区办事处，以及所有车队管理中心和下属单位。
Interchange: An area designated to provide traffic access between roadways of differing levels.	**立交**：为不同层次之间的道路的交通提供连接的区域。
Intercity and Rural Bus Transportation: Establishments primarily engaged in furnishing bus transportation, over regular routes and on regular schedules, the operations of which are principally outside a single municipality and its suburban areas.	**城际和农村公共汽车交通**：主要在城市及其郊区外围运营，并有固定路线及时刻表的公共交通设施。
Intercity Bus: A standard size bus equipped with front doors only, high backed seats, luggage compartments separate from the passenger compartment and usually with rest room facilities, for high-speed long distance service.	**城际汽车**：标准尺寸的公共汽车，只有前门，有高靠背座椅，与乘客车厢分离的行李舱，通常有卫生间设施，为高速长途旅行服务。

English	Chinese
Intercity Rail Passenger: A rail car, intended for use by revenue passengers, obtained by the National Railroad Passenger Corporation (currently AMTRAK) for use in intercity rail transportation.	**城际铁路客运**：一个轨道车，供普通乘客使用，由全国铁路客运公司获得的（AMTRAK），在城际轨道交通使用。
Intercity Rail Transportation: Transportation provided by Amtrak.	**城际轨道交通**：由美铁提供的交通。
Intercity Transportation: 1) Transportation between cities. 2) Transportation service provided between cities by certified carriers, usually on a fixed route with a fixed schedule.	**城际交通**：1）城市之间的交通。2）由执照运营商提供的城市之间的服务，通常在一个固定的时间表和固定的路线。
Intercity Trucking: Trucking operations which carry freight beyond the local areas and commercial zones.	**城际货运**：在商业区和本地区以外运营的货物运输。
Interest Long Term Debt and Capital Leases: Interest on all classes of debt, both short-term and long-term, as well as the amortization of premium, discount and expense connected with the issuance of such debt and interest expense on capital leases.	**利息长期债务与资本租赁**：由各种债务获取的利息，包括短期和长期的，和与这些债务相关的折旧，优惠和消费，以及资本租赁的利息费用。
Intergovernmental Revenue: Amounts received from other governments as fiscal aid in the form of shared revenues and grants-in-aid, as reimbursements for performance of general government funct-ions and specific services for the paying government, or in lieu of taxes. This revenue excludes amounts received from other governments for sale of property, com-modities and utility services.	**政府间收入**：从其他政府得到的收入，包括共享收入和赠款形式的财政援助，，执行一般政府职能和具体服务的支付，或用税收代替偿还款项。这项收入不包括向其他政府出售财产，商品和公用事业服务收到的款项。

English	Chinese
Interior Compartment Door: Any door in the interior of the vehicle installed by the manufacturer as a cover for storage space normally used for personal effects.	**车厢内门**：任何由汽车制造商安装在车辆内部作为存储空间的门，通常用于保存个人财物。
Interlocked Route: A route within interlocking limits.	**联锁路线**：连锁范围内的路线。
Interlocked Switch: A switch within the interlocking limits the control of which is interlocked with other functions of the interlocking.	**联锁开关**：联锁的开关限制了它与其他的联锁控制。
Interlocking Limits: The tracks between the opposing home signals of an interlocking.	**联锁限制**：相对于连锁装置的进站信号之间的路线。
Interlocking Machine: An assemblage of manually operated levers or other devices for the control of signals, switches or other units.	**联锁机**：一些手动杠杆或其它设备的组合，以控制信号，开关或其他设备。
Interlocking Signal: A roadway signal which governs movements into or within interlocking limits.	**联锁信号**：一种路面信号以用来管理进入或在联锁限制范围内的运动。
Intermediate Grade Gasoline: An increasingly common grade of unleaded gasoline with an octane rating intermediate between "regular" and "premium". Octane boosters are added to gasolines to control engine pre-ignition or "knocking" by slowing combustion rates.	**中级汽油**：一种越来越普遍的辛烷值在"常规"及"高级"之间的无铅汽油。辛烷值助推器被添加到汽油中，通过让燃烧率变慢来控制发动机提前点火或爆震。
Intermediate Type Road Surface: Mixed bituminous and bituminous penetration (Surface/Pavement Type Codes 52 and 53).	**中间型路面**：沥青及沥青混合渗透（表面/路面类型代码 52 和 53）。

English

Intermittent, Casual, or Occasional Driver: A driver who in any period of 7 consecutive days is employed or used as a driver by more than a single motor carrier. The qualification of such a driver shall be determined and recorded in accordance with the provisions of 49 CFR 391.63 or 391.65 as applicable.

Intermodal Container: A freight container designed and constructed to permit it to be used interchangeably in two or more modes of transport.

Intermodal Passenger Terminal: An existing railroad passenger terminal which has been or may be modified as necessary to accommodate several modes of transportation, including intercity rail service and some or all of the following: intercity bus, commuter rail, intra-city rail transit and bus transportation, airport limousine service and airline ticket offices, rent-a-car facilities, taxis, private parking, and other transportation services.

Intermodal Portable Tank: A specific class of portable tanks designed primarily for international intermodal use.

Intermodalism: Typically used in three contexts: 1) most narrowly, it refers to containerization, piggyback service, or other technologies that provide the seamless movement of good and people by more than

Chinese

间歇性，偶然，或不定期司机：在任何连续 7 天时间内被一个以上的汽车运输业的公司雇用的司机。这种司机的资格，应按照 49CFR391.63 或 391.65 的规定决定和记录。

多式联运集装箱：一个货物集装箱的设计及建造，允许它使用两种或多种运输方式间交替。

联运客运大楼：一个现有的铁路客运码头已经或可能需要改造，以容纳多种运输方式，包括城际铁路和如下的一些或全部：城际巴士，通勤铁路，市内轨道交通和公共汽车交通运输，机场接送服务，航空公司售票处，租用汽车的设施，的士，私家停车场等运输服务。

多式联运便携式体：一个主要用于国际联运而设计的便携式罐体。

多式联运：通常在三个方面：1）最狭义的，是指集装箱，搭载服务，或其他技术为货物或乘客提供一个以上的运输方式。2）较广义，多式联运是

English

one mode of transport. 2) more broadly, intermodalism refers to the provision of connections between different modes, such as adequate highways to ports or bus feeder services to rail transit. 3) In its broadest interpretation, intermodalism refers to a holistic view of transportation in which individual modes work together or within their own niches to provide the user with the best choices of service, and in which the consequences on all modes of policies for a single mode are considered. This view has been called balanced, integrated, or comprehensive transportation in the past.

Internal Combustion Engine: An engine in which the power is developed through the expansive force of fuel that is fired or discharged within a closed chamber or cylinder.

Interstate: Limited access divided facility of at least four lanes designated by the Federal Highway Administration as part of the Interstate System.

Interstate Commerce: Trade, traffic, or transportation in the United States which is between a place in a State and a place outside of such State (including a place outside of the United States) or is between two places in a State through another State or a place outside of the United States.

Chinese

指为不同模式提供连接，譬如提供足够的公路到港口，或巴士服务与轨道连接。 3）最广义的解释，多式联运是指一种交通观念，其中各个运输方式一起或在自己最合适的位置工作，为用户提供服务的最佳选择，并且一种模式的政策对所有模式的后果得到考虑。这种观念在过去被叫做平衡，综合，全面的交通。

内燃机：借助燃料在封闭室或气缸发动机中引燃导致的发射膨胀力而获得力量的引擎。

州际公路：联邦公路管理局指定的作为州际系统一部分的，限制入口、双向隔离、至少四车道的设施。

州际贸易：在美国的一个州和州外（包括境外）之间的贸易，交通，或运输，或通过另一个州或美国以外的地方实现一个州内的两个地方之间的交易。。

English	Chinese
Interstate Commerce Commission (ICC): The federal body charged with enforcing Acts of Congress affecting interstate commerce.	州际商务委员会（ICC）：执行影响州际商业行为的国会法案的联邦机构。
Interstate Commerce Commission (ICC) Authorized Carrier: A for-hire motor carrier engaged in interstate or foreign commerce, subject to economic regulation by the ICC.	州际商务委员会（ICC）指定运营商：被雇用从事州际或国外的商业活动的乘运人，须经**州际商务委员会**经济法规的管理。
Interstate Commerce Commission (ICC) Exempt Carrier: A for-hire motor carrier transporting commodities or conducting operations not subject to economic regulation by the ICC.	州际商务委员会（ICC）豁免运营商：被雇用从事州际或国外商业活动的乘运人，不受**州际商务委员会**经济法规的管理。
Interstate Highway (Freeway or Expressway): (See also Expressway, Freeway, Freeways and Expressways, Highway) A divided arterial highway for through traffic with full or partial control of access and grade separations at major intersections.	州际公路（高速公路或快速公路）：（又见快速公路，高速公路，高速公路和快速公路，公路）双向分离的、用于通过交通的、有全部或部分入口限制权的、并在交叉路口有等级分离的干线公路。
Interstate Highway System: This system is part of the Federal Aid Primary system. It is a system of freeways connecting and serving the principal cities of the continental United States.	州际公路系统：本系统是联邦主要援助系统的一部分。它是连接和服务于美国本土的主要城市的高速公路系统。
Interstate Pipeline: A pipeline or that part of a pipeline that is used in the transportation of hazardous liquids or carbon dioxide in interstate or foreign commerce.	州际管道：一个管道或是一个管道的一部分用来在州际或国际之间运输危险液体或二氧化碳。

English	Chinese
Intrastate: Travel within the same state.	**州内**：在一个州内的旅行。
Intrastate Commerce: Any trade, traffic, or transportation in any State which is not described in the term "interstate commerce."	**州内贸易**：任何州内贸易，交通，或运输，并未在"州际贸易"中描述到。
Intrastate Pipeline: See also Interstate Pipeline.	**州内的管道**：参见州际管道。
Intrastate Pipeline: A natural gas pipeline company engaged in the transportation, by pipeline, of natural gas not subject to the jurisdiction of the Federal Energy Regulatory Commission (FERC) under the Natural Gas Act.	**州内的管道**：通过管道从事天然气运输的天然气管道公司，，根据天然气法不受联邦能源管制委员会的管辖范围（联邦能源管制委员会）。
Investments and Special Funds: Investments and advances to investor controlled and other associated companies, notes and receivables not due within one year, investment in securities issued by others, allowance for unrealized gain or loss on noncurrent marketable equity securities, funds not available for current operations, investments in leveraged leases, and net investments in direct financing and sales-type leases which are not reasonably expected to be amortized within one year.	**投资基金和特别基金**：受投资人控制及其相关公司的投资和利益，不在一年内到期的票据和应收账款，证券投资，非流动销售股本证券中未实现收益或损失津贴，不能用于目前的业务的资金，杠杆租赁投资，不在一年内折旧的净投资和和销售型租赁。
Iron: Old model truck.	**铁车**：旧型卡车。

English	Chinese
Joint Operations: Rail operations conducted on a track used jointly or in common by two or more railroads subject to 49 CFR 225 or operation of a train, locomotive, car or other on-track equipment by one railroad over the track of another railroad.	**联合操作**：两家或两家以上铁路公司 遵照 49 CFR 225 在同一個轨道上进行铁路运输或一家铁路公司在另一家公司的轨道上运送车辆，机车，汽车或其他轨道上的操作设备。
Jumped the Pin: Missing the fifth wheel pin on the trailer when coupling tractor to trailer.	**跳杆**: 挂车斗时错过第五个车轮的插头
Junction: Area formed by the connection of two roadways, including intersections, interchange areas and entrance/exit ramps.	**交界处**：两个车道连接区，包括交叉路口，交换区和入口/出口斜道的连接。

English

Kerosene: (See also Fuel, Gasoline) A petroleum distillate that boils at a temperature between 300 and 550 degrees Fahrenheit, that has a flash point higher than 100 degrees Fahrenheit by ASTM Method D 56, that has a gravity range from 40 to 46 degrees API, and that has a burning point in the range of 150 degrees to 175 degrees Fahrenheit. Kerosene is used in space heaters, cook stoves, and water heaters and is suitable for use as an illuminate when burned in wick lamps.

King-Pin Saddle-Mount: (See also Lower Half of Saddle-Mount, Saddle-Mount, and Upper Half of Saddle-Mount) That device which is used to connect the "upper-half" to the "lower-half" [of a " saddle-mount"] in such manner as to permit relative movement in a horizontal plane between the towed and towing vehicles.

Chinese

煤油：（另见燃料，汽油）石油馏分的沸点 300 至 550 华氏的温度，具有闪点高 于 100 度 由 ASTM 方法 D 56，具有美国石油协会 40 至 46 度重力指标，燃烧点在 150 度至 175 华氏度。煤油用于空间加热器，厨灶，热水器，并使用照亮，使用灯芯灯。

竞品鞍山：（也见的下半鞍山，马鞍山，和上鞍半山），这是用来连接"上半"到"下半"的设备一个"马鞍安装"的方式]，允许牵引和拖车间水平面间的相对运动。

English	Chinese

Land Area: Based on the U.S. Bureau of the Census definition, this includes dry land and land temporarily or partially covered by water, such as marshlands, swamps and river flood plains, systems, sloughs, estuaries and canals less than 1/8 of a statute mile (0.2 kilometers) in width and lakes, reservoirs and ponds less than 1/16 square mile (0.16 square kilometers) in area. [For Alaska, 1/2 mile (0.8 kilometers) and 1 square mile (2.60 square kilometers) are substituted for these values]. The net land area excludes areas of oceans, bays, sounds, etc., lying within the 3 mile (4.8 kilometers) U.S. jurisdiction as well as inland water areas larger than indicated above.

国土面积：根据美国人口普查局的定义，包括干陆地和临时或部分被水覆盖的陆地，如湿地、沼泽和河流冲击平原、水系、泥潭、河口及宽度少于 1/8 法定英里(0.2 公里)的运河，面积不到 1/16 平方英里(0.16 平方公里)的湖泊、水库、池塘。[对于阿拉斯加，用 1/2 英里(0.8 公里)，一平方英里(2.60 平方公里)来代替这些数据]。土地净面积指除了海洋,海湾,海峡等面积,离美国境内三英里(4.8 公里)区域内的国土以及大于上述面积的内陆水域。

Land Use: Designates whether the general area in which the crash occurred is urban or rural, based on 1980 Census Data.

土地利用：指基于 1980 年普查数据，确定发生交通事故的地区被归为城市还是乡村地区。

Landing Gear: Device that supports the front end of semitrailer when not attached to a tractor.

支撑架: 在没有卡车牵引车头时，支撑双轮拖车前端的装置。

Landscaping: (See also Brush Out) Colloquial term meaning to clear shore structure of brush and vegetation in order to obtain optimum range of visibility.

景观美化：（又见"清除"）指清除河岸边的结构物及植被以获得最适宜范围的视线的通俗术语。

Lane: A prescribed course for ships or aircraft, or a strip delineated on a road to accommodate a single line of automobiles; not to be confused with the road itself.

专用道：为船舶、飞机所指订的的专用通道，或者是在公路上划出的专供汽车行驶车道线，车道与道路本身不应混淆。

English	Chinese
Lane: A portion of a street or highway, usually indicated by pavement markings, that is intended for one line of vehicles.	**车道线**：街道或公路的一部分，通常由路面标线来表示，主要用于规范车辆在一条线上行驶。
Large Truck: Trucks over 10,000 pounds gross vehicle weight rating, including single unit trucks and truck tractors.	**大型卡车**:包括单个卡车和有牵引拖车的卡车等总重量超过一万磅的卡车。
Latch Block: The lower extremity of a latch rod which engages with a square shoulder of the segment or quadrant to hold the lever in position.	**制动闸**：和一个方形的装置或四分仪相啮合，保持控制杆位置的制动销的最底端。
Latch Shoe: The casting by means of which the latch rod and the latch block are held to a lever of a mechanical interlocking machine.	**制动鞋**：制动杆和制动销通过其连接到机械互锁杠杆的铸件装置。
Lay On the Air: Apply brakes.	**空气制动**：用于制动。
Layover: Eight hours or more rest before continuing trip or any off-duty period away from home.	**临时停靠**：休息八小时或以上，然后继续行程或任何在外的非工作时间。
Leaded Motor Gasoline: (See also Gasoline) Contains more than 0.05 grams of lead per gallon or more than 0.005 grams of phosphorus per gallon. The actual lead content of any given gallon may vary. Premium and regular grades are included, depending on the octane rating also leaded gasohol. Blendstock is excluded until blending has been completed. Alcohol that is to be used in the blending of gasohol is also excluded.	**含铅汽油**: (又见汽油)每加仑含铅超过 0.05 克或含磷 0.005 克以上的汽油。任何给定的每加仑汽油实际含铅量可能不一样。包括高级和正常等级的汽油，主要根据其辛烷值来确定其是否含铅汽油，不包括混合汽油。直到完全融合混合成分才被排除。也不包括用于勾兑乙醇汽油的酒精。

English	Chinese
Lease: Acquisition of a vehicle by an agency from a commercial firm, in lieu of government ownership, for a period of 60 continuous days or more.	**租赁**：一政府部门通过从商业公司租赁汽车，并连续租借 60 天或以上，以取代该部门拥有汽车。
Leased Property (Under Capital Leases): Total cost to the air carrier for all property obtained under leases that meet one or more of the following criteria; 1) The lease transfers ownership of the property to the lessee by the end of the lease term; 2) The lease contains a bargain purchase option; 3) The lease term is equal to 75 percent or more of the estimated economic life of the leased property; or 4) The present value at the beginning of the lease term of the minimum lease payments, excluding the portion of the payments representing executory costs such as insurance, maintenance and taxes to be paid by the lesser, including any profit thereon, equals or exceeds 90 percent of the excess of the fair value of the lease property to the lesser at the inception of the lease over any related investment tax credit retained by the lesser and expected to be realized by him.	**租赁财产(资本租赁)**：对于航空运输商在符合以下一个或多个标准的租赁契约条件下获得的所有财产所支付的总成本：1) 租期结束时租赁财产所有权转让给承租人；2)租赁包含一个契约购买合同；3)租期要相当于 75%或以上的的出租设备的估计使用期限；4)在最低租赁费的租期开始时的净现值，不包括尚未花销的费用如：保险、维修以及出租人需要支付的税款，包括任何后期利润，达到或超过 90%的租赁财产的过剩公允价值出租人在租赁起初获得的任何投资税款。
Less Than Truckload (LTL): A quantity of freight less than that required for the application of a truckload rate. Usually less than 10,000 pounds and generally involves the use of terminal facilities to break and consolidate shipments.	**散货**：需要运送的货物不足一卡车的额定容量。一般少于一万磅,通常需要使用港口设施来分装和组合运输货柜。

English	Chinese
Level of Service: 1) A set of characteristics that indicate the quality and quantity of transportation service provided, including characteristics that are quantifiable and those that are difficult to quantify. 2) For highway systems, a qualitative rating of the effectiveness of a highway or highway facility in serving traffic, in terms of operating conditions. 3) For paratransit, a variety of measures meant to denote the quality of service provided, generally in terms of total travel time or a specific component of total travel time. 4) For pedestrians, sets of area occupancy classifications to connect the design of pedestrian facilities with levels of service.	**服务水平**:1)一组表示所提供的运输服务的质量和数量的特征,包括可量化的和很难量化的特性。2)在运营状况方面，对于公路系统而言,对公路及公路设施在交通服务上的效用进行的非量化评估。3)对于辅助公交系统,各种指标表示提供服务的质量,一般为在整体旅行时间或其中特定的一部分等方面。4)对于行人，人行道的占用面积的布置和服务水平的人行道设施的设计相联系。
License Plate Lamp: A lamp used to illuminate the license plate on the rear of a motor vehicle.	**车牌灯**：用来照亮汽车尾部车牌照的灯。
Licensed Driver: Any person who holds a valid driver's license from any state.	**有执照司机**：持有任何州颁发的有效驾照的司机。
Lie Sheet: Driver's log book.	**记录表**：(卡车)司机的日常行程记录。
Light Density Railroad: Railroads with 1200 or less train-miles per road mile.	**低密度铁路**：以每一道路英里为单位,铁路列车行驶低于一千二百火车英哩或者更低的铁路系统。
Light Duty Vehicle: Automobiles and light trucks combined.	**轻型车**：汽车和轻型卡车的总和。

English

Light Truck: An automobile other than a passenger automobile which is either designed for off-highway operation as described in the second sentence below or designed to perform at least one of the following functions: 1) Transport more than 10 persons; 2) Provide temporary living quarters; 3) Transport property on an open bed; 4) Provide greater cargo-carrying than passenger-carrying volume; or 5) Permit expanded use of the automobile for cargo-carrying purposes or other nonpassenger-carrying purposes through the removal of seats by means installed for that purpose by the automobile's manufacturer or with simple tools, such as a screwdriver and wrenches, so as to create a flat, floor level, surface extending from the forward most point of installation of those seats to the rear of the automobile's interior. An automobile capable of off-highway operation is an automobile a) That has 4-wheel drive; or is rated at more than 6,000 pounds gross vehicle weight; and b) That has at least four of the following characteristics calculated when the automobile is at curb weight, on a level surface, with the front wheels parallel to the automobile's longitudinal centerline, and the tires inflated to the manufacturer's recommended pressure: (i) Approach angle of not less than 28 degrees; (ii) Breakover angle of not less than 14 degrees; (iii) Departure angle of not less than 20 degrees; (iv) Running clearance of not less than 20

Chinese

轻型卡车: 一种除具有能搭载乘客以外还可以设计成如下面第二句描述的那样可越野运行或者设计成至少具有下列一个职能的汽车: 1)可载客超过 10 人; 2)提供临时居住空间; 3)运输物品可置于敞开车板上; 4)提供比载客量更大的货物运载量; 或者 5)具有特许证并通过由汽车制造商或自己用简单工具如螺丝刀、扳手等来拆除座位,制造一个水平车地板。表面扩展从作为安装座位的最前沿到汽车内部的后面从而取消座位的固定位置来使汽车达到扩大运载货物或者其他非乘客的用途。 能够在野外营运的汽车:a)四轮驱动; 或者是其额定车辆总重超过六千磅; 并且 b)当汽车处于静止在水平地面,与前面的汽车车轮平行的纵向中线上,在充气轮胎的制造商建议的压力时至少具有以下四个计算特征:(i)接近角度不少于 28 度; (ii)转折角度不少于 14 度; (iii)离开角度不少于 20 度; (iv)旋转间隙不少于 20 厘米; (v)前后轴间隙不少于 18 公分以上。

English	Chinese

centimeters; (v) Front and rear axle clearances of not less than 18 centimeters each.

Light Truck: Trucks of 10,000 pounds gross vehicle weight rating or less, including pickups, vans, truck-based station wagons, and utility vehicles.

轻型卡车: 其额定车重不超过一万磅的卡车,包括皮卡、面包车、以卡车为基础的无后箱轿车和吉普车辆。

Light Truck: Two-axle, four-tire trucks.

轻型卡车：两轴四轮的卡车。

Limousine or Auto Rental With Driver: Establishments primarily engaged in furnishing limousines or auto rentals with drivers, where such operations are principally within a single municipality, contiguous municipalities, or a municipality and its suburban areas eg. automobile rental with driver, limousine rental with driver, hearse rental with driver, passenger automobile rental with driver.

豪华出租轿车或带司机的出租车：主要从事装饰豪华轿车或带出租司机的汽车出租业务。这类业务主要是在一个市,毗邻市或者是城市及其郊区等等。如：连带司机的出租汽车,出租司机的轿车、出租司机的灵车和出租司机的客车。

Line: One or more running tracks, each kilometre of line counting as one, however many tracks there may be. The total length of line operated is the length operated for passenger or goods transport, or both. Where a section of network comprises two or more lines running alongside on another, there are as many lines as routes to which tracks are allotted exclusively.

线路:一条或多条运行线路,每公里线路计作一个单位,可能会有很多线路。可运行的线路的总长度是用于客运或货运经营,或者两者兼有的线路。凡在由两个或两个以上运行的紧挨的网络线路的地方,可能有同样多的线路但轨道是独立的。

Line-Haul: Transportation from one city to another as differentiated from local switching service.

长途运输: 和本地运输服务相区分的从一个城市到另一个城市的运输。

English	Chinese

Line Miles: The sum of the actual physical length (measured in only one direction) of all streets, highways, or rights-of-way traversed by a transportation system (including exclusive rights-of-way and specially controlled facilities) regardless of the number of routes or vehicles that pass over any of the sections.

线路里程: 不计路径数量和其上运行车辆的数目的实际线路长度的总和(只在一个方向运行)。包括所有街道、公路、运输系统横贯的(包括专属可通行里程及特别控制的交通设施)的道路。

Linear Referencing System (LRS): The total set of procedures for determining and retaining a record of specific points along a highway. Typical systems used are mile point, milepost, reference point, and link-node.

线性参照系统: 用来确定和记录公路沿线具体位置的一整套程序。采用的典型系统有里程点、里程桩、参考点和线路节点法。

Linked Passenger Trip: A trip from origin to destination on the transit system. Even if a passenger must make several transfers during a journey, the trip is counted as one linked trip on the system.

连续旅程: 在公交系统中从起始点到目的地的一段路程。即使是乘客在途中必须换乘几次,也算作是一段连续的行程。

Lives Lost: Those persons who perished as a direct result of the distress incident to which the Coast Guard was responding. Lives Lost Before refers to those persons who were considered lost prior to Coast Guard notification. Lives Lost After refers to those persons who were alive at the time of Coast Guard notification, but who subsequently died.

丧失的生命: 由海岸警卫队负责救的惨痛事件的直接受害人。提前丧生是指那些在海岸警卫队通知前失去生命的人,滞后丧生是指那些在海岸警卫队通知时还活着但后来死去的人。

Lives Saved: Those persons who would have been lost without Coast Guard assistance.

拯救的生命:那些没有海岸警卫队的援助将会失去生命的人。

English	Chinese
Livestock Body: Truck or trailer designed for the transportation of farm animals.	**畜牧运输车**：用来运载家畜的卡车或拖车。
Load Factor: The percentage of seating or freight capacity which is utilized.	**乘载系数**: 乘客与座位及货载与货物运输容量的百分比。
Load Factor: A term relating the potential capacity of a system relative to its actual performance. Is often calculated as total passenger miles divided by total vehicle miles.	**负荷因子**：和其实际业绩相对的的承载能力相关的术语。常常用总运送乘客里程除以总车英里来计算。
Loaded Car Mile: A loaded car mile is a mile run by a freight car with a load. In the case of intermodal movements, the car miles generated will be loaded or empty depending on whether the trailers/containers are moved with or without a waybill, respectively.	**装载英里**：一个装载英里是货车装载时运行一英里。在联合运输的过程中，产生的装载英里将视挂车/集装箱是否有货单而定是装载或是空载。
Loading Island: 1) A pedestrian refuge within the right-of-way and traffic lanes of a highway or street. It is provided at designated transit stops for the protection of passengers from traffic while they wait for and board or alight from transit vehicles. 2) A protected spot for the loading and unloading of passengers.	**车站岛**：1)在可通行的线路和公路或街道的交通路线之间的一个为行人设置的安全场所。在指定线路车站，它可以使过境旅客远离过往车辆；2)供乘客上下车的保护场所。
Local and Suburban and Interurban Passenger Transportation Transit: Includes establishments that provide local and suburban passenger transportation, such as those providing passenger transportation within a single municipality, contiguous municipalities, or a municipality and its suburban areas by bus, rail, car subway, either separately or in combination.	**本地和市郊及城市间客运公交**：包括提供本地及近郊客运，如在本市、毗邻的城市、城市及其郊区等的巴士、铁路和地铁，单独或混合乘坐的公司。还包括观光，出租,城际客运业务,并设有客运及维修设施的公司。

English	Chinese

Also included are sightseeing, charter, intercity passenger operations, and establishments providing passenger terminal and maintenance facilities.

Local and Suburban Transit: Establishments primarily engaged in furnishing local and suburban mass passenger transportation over regular routes and on regular schedules, with operations confined principally to a municipality, contiguous municipalities, or a municipality and its suburban areas. Also included in this industry are establishments primarily engaged in furnishing passenger trans-portation by automobile, bus, or rail to, from, or between airports or rail terminals over regular routes and those providing bus and rail commuter services.

本地和市郊公交：主要从事本地及市郊客运定期定线和正规的班次服务。业务主要局限于本市, 毗邻市政府、市及其郊区的设施。也包括主要从事客运的汽车、公共汽车或火车, 到、来自或在机场或铁路终点站之间固定线路及提供巴士和铁路通勤服务的公司。

Local Courier Service: Establishments primarily engaged in the delivery of individually addressed letters, parcels, or packages (generally under 100 pounds), except by means of air transportation or by the United States Postal Service. Delivery is usually made by street or highway within a local area or between cities.

本地速递业务: 除了借由航空运输或美国邮政服务以外，主要从事单独信件、包裹(一般在一百磅以下)的运送等业务的公司。速递业务通常是借助当地或者是城市之间的街道或者公路来完成的。

Local Freight: Freight movements within the confines of a port, whether the port has only one or several arms or channels (except car ferry and general ferry). The term is also applied to marine products, sand, and gravel taken directly from the Great Lakes.

本地货运：货物在港口内运输, 不管是否只有一个或几个港口分支或通路(除汽车客运及一般码头外)。这个术语也适用于海产品、沙子和直接从五大湖运输的砾石。

English	Chinese
Local Passenger (Not Elsewhere Classified) Transportation: Establishments primarily engaged in furnishing miscellaneous passenger transportation, where such operations are principally within a single municipality, contiguous muni-cipalities, or a municipality and its suburban areas.	**本地乘客运输 (没有其他分类)**: 主要从事各种客运，在本市、邻市、及其郊区运行的公司。
Local Roads: Those roads and streets whose principal function is to provide direct access to abutting land.	**地方道路**: 其主要功能是提供直接进出毗邻土地的道路和街道。
Local Streets and Roads: (See also Highway, Road) Streets whose primary purpose is feeding higher order systems, providing direct access with little or no through traffic.	**地方街道和道路**: (又见公路,道路),主要用途是向高等级公路提供交通流，很少或没有过境交通。
Local Trip: An intracity or short mileage trip by commercial motor vehicle.	**本地旅程**：乘坐商用客车的市内或短途的旅程。
Local Trucking (With Storage): Establish-ments primarily engaged in furnishing both trucking and storage services, including household goods, within a single municipality, contiguous municipalities, or a municipality and its suburban areas.	**本地卡车货物运输 (带仓库)**：主要从事运输及贮存的包括日用品的服务,在本城市、毗邻的城市，或城市及其郊区之间运行的运输公司。
Local Trucking (Without Storage): Estab-lishments primarily engaged in furnishing trucking or transfer services without storage for freight generally weighing more than 100 pounds, in a single municipality, contiguous municipalities, or a municipality and its suburban areas.	**当地卡车货物运输 (无仓库)**: 主要从事一般不储存超过一百磅的货物的货运业务或中转服务，主要在本城市、毗邻的城市，或城市及其郊区之间运行的运输公司。

English

Lock Rod: [with respect to rail operations] A rod, attached to the front rod or lug of a switch, movable-point frog or derail, through which a locking plunger may extend when the switch points or derail are in the normal or reverse position.

Locking Bar: [with respect to rail operations] A bar in an interlocking machine to which the locking dogs are attached.

Locking Bed: [with respect to rail operations] That part of an interlocking machine that contains or holds the tappets, locking bars, crosslocking, dogs and other apparatus used to interlock the levers.

Locking Dog: (See also Dog Chart) [with respect to rail operations] A steel block attached to a locking bar or tappet of an interlocking machine, by means of which locking between levers is accomplished.

Locking Face: [with respect to rail operations] The locking surface of a locking dog, tappet or cross locking of an interlocking machine.

Locking Sheet: A description in tabular form of the locking operations in an interlocking machine.

Locking Time: The total time required for a tow to pass through a locking procedure. This includes approach time, chamber time, and time to clear the lock.

Chinese

闸门杆 [用于轨道运行]：一根连接到前杆或者是开关接线片的、活动的点状辙叉的或脱轨器上的连杆，通过它当开关点或脱轨器处于正常或相反位置时闸门开关柱塞可以延长。

闸门杆 [用于轨道运行]：在一个附有闸门的联锁的机器上的连杆。

闸门床 [用于轨道运行]：联锁机器的一部分，包含或者控制凸轮推杆、抱死杆、交叉制动、锁定狗和其他用于互锁控制杆的器具。

闸门搭扣 (参看搭扣图表)[用于轨道运行]：一个连接在锁定杆或联锁机器的凸轮推杆上的钢块，用它完成控制杆之间的互锁。

锁面 [用于轨道运行]：闸门搭扣的锁面，凸轮推杆或者是互锁的联锁机的表面。

锁面表：以表格形式描述联锁机的锁定业务。

锁定时间：缆绳通过抱死过程所需要的总时间。包括接近时间、穿过时间和离开时间。

English	Chinese
Locomotive: A self-propelled unit of equipment designed primarily for moving other equipment. It does not include self-propelled passenger cars.	**机车**：一个自我驱动用来牵引其他设备的机动式设备单元，不包括自我驱动机动客车。
Locomotive Mile: The movement of a locomotive under its own power the distance of one mile.	**机车英里**：一个机车单元靠自身动力运行一英里的距离。
Locomotive Unit Mile: The movement of a locomotive unit one mile under its own power. Miles of locomotives in helper service are computed on the basis of actual distance run in such service. Locomotive unit miles in road service are based on the actual distance run between terminals and/or stations. Train switching locomotive unit miles are computed at the rate of six miles per hour for the time actually engaged in such service.	**机车单位英里**：一个机车单元靠自身动力运动的一英里。有助手服务的机车里程的计算要以其实际运行长度为基础。在路途业务中的机车单元里程是基于起、终点和/或车站之间的实际运行长度。机车配电系统单元里程的计算是在从事这种业务的时间达到每小时六英里的速率。
Log Body: Truck or trailer designed for the transportation of logs or other loads which may be boomed or chained in place.	**原木机车**：用来运送原木或其它的要用锁链固定的负载的卡车或拖车。
Longer Combination Vehicles (LCV): Any combination of truck tractor and two or more trailers or semitrailers which operates on the Interstate System at a gross vehicle weight greater than 80,000 pounds.	**加长结合车**: 在州际高速公路运行的车辆总重超过八万磅的卡车、两个或多个拖车、半拖车的任意组合。
Longitudinal: Parallel to the longitudinal centerline of the vehicle.	**纵轴**：纵向平行于汽车的纵向中心线。
Lost Workdays: Any full day or part of a day (consecutive or not) other than the day of	**失去的工作日**: 除了受工伤的日子的任何全天或任何一天的

English	Chinese
injury, that a railroad employee is away from work because of injury or occupational illness.	一部分 (连续或不连续的)，如一个铁路雇员因工受伤或职业病而不能工作。
Low Boy: A low trailer for hauling heavy machinery.	**低拖车**：承载重型机械的低拖车。
Low Emission Vehicle: A clean fuel vehicle meeting the low-emission vehicle standards.	**低排放车辆**: 达到低排放车辆标准的清洁燃料车。
Low Type Road Surface: Bituminous surface-treated Surface/Pavement Type Code	**低类型路面**：表面经沥青处理过的路面。代码 51。
Lower-Half of Saddle-Mount: (See also King-Pin Saddle-Mount, Saddle-Mount, Upper-Half of Saddle-Mount) That part of the device which is securely attached to the towing vehicle and maintains a fixed position relative thereto but does not include the "king-pin."	**下半鞍座**：又见主销鞍座，鞍座，上半鞍座，安全地连接到牵引车上并保持相对固定位置，但不包括"主销"的装置的一部分。
Lumber Body: Platform truck or trailer body with traverse rollers designed for the transportation of sawed lumber.	**木材机车**：用来运送锯好的木头的带有滚轴的平台卡车或拖车。

English	Chinese
Mail Revenue: Revenues from the carriage of mail bearing postage for air transportation both U.S. and foreign mail that go by air on priority and nonpriority bases.	**邮政运输收入**：来自邮件的运费，这些运费是美国和其它国家通过具有优先权和非优先权的航空运输的邮件所获取来的。
Main Heating Fuel: Fuel that powers the main heating equipment.	**主加热燃料**：为主要加热设备提供动力的燃料。
Maintenance: All expenses, both direct and indirect, specifically identifiable with the repair and upkeep of property and equipment.	**维护费用**：包括直接和间接的一切维护费用，尤指财产和设备的维修费。
Maintenance Control Center (MCC): Responsible for the oversight of authorization for vehicle repair and authorization and certification of maintenance and repair invoices for Interagency Fleet Management System (IFMS) vehicles within the specified region(s). The MCC also contacts vendors to schedule vehicle services.	**维护控制中心（MCC）**：负责监管车辆维修的授权以及对机构内部快速管理系统（IFMS）车辆的维修和维护予以授权和证明。MMC 还联系卖主制定车辆服务的时间表。
Major Fuel: Fuels or energy sources such as electricity, fuel oil, liquefied petroleum gases, natural gas, district steam, district hot water, and district chilled water.	**主燃料**：燃料或能量资源，如：电力，燃油，液化气，天然气，以及区域内的蒸汽，热水和冰水。
Mandatory Use Seat Belt Law: A law requiring some adult occupants of some traffic vehicles to use available restraint systems.	**强制使用安全带法规**：本法规要求拥有交通车辆的成年人使用现有的约束系统。
Manner of Collision: A classification for crashes in which the first harmful event was	**碰撞类别**：碰撞分类，这些碰撞中最初的伤害是行驶中两辆

English	Chinese
a collision between two motor vehicles in transport.	机动车辆间的碰撞。

Manual Interlocking: An arrangement of signals and signal appliances operated from an interlocking machine and so interconnected by means of mechanical and/or electric locking that their movements must succeed each other in proper sequence, train movements over all routes being governed by signal indication.

手动互锁：信号和信号装置的排列。该信号装置由一种机械和/或电子联锁器控制，它们按一定顺序运动，使所有路线上的运动按信号指示操作。

Manual Restraint System: (See also Mandatory Use Seat Belt Law, Restraint Usage) Occupant restraints that require some action, usually buckling, before they are effective. They include shoulder belt, lap belt, lap and shoulder belt, infant carrier, or child safety seat.

手动约束系统：操作后才产生效用的约束系统，通常为扣住。它们包括肩带，腰带，肩和腰都束缚的安全带，婴儿车或小孩安全座。

Marine Cargo Handling: Establishments primarily engaged in activities directly related to marine cargo handling from the time cargo, for or from a vessel, arrives at shipside, dock, pier, terminal, staging area, or in-transit area until cargo loading or unloading operations are completed. Included in this industry are establishments primarily engaged in the transfer of cargo between ship and barges, trucks, trains, pipelines, and wharfs. Cargo handling operations carried on by transportation companies and separately reported are classified here. This industry includes the operation and maintenance of piers,

海运货物装卸：与海运货物装卸直接相关的一系列操作，包括实时货物运抵码头、集结地或中转地、终点，直至货物装载或卸载完毕。业内公司主要包括涉及货物在船和码头、卡车、火车、管道之间运输的公司。货物装卸操作由运输公司完成并分别报告。这个行业包括码头以及相关建筑和设施的维护和运作。

English	Chinese

docks, and associated buildings and facilities.

Marking: A descriptive name, identification number, instructions, cautions, weight, specification, or combinations thereof, required by this subchapter on outer packagings of hazardous materials.

标记：在危险物品的外包装上要求标记的一个描述性的术语，可以是鉴定数码、说明、警告、重量、类别或它们的组合。

Maxi-Cube Vehicle: A combination vehicle consisting of a power unit and a trailing unit, both of which are designed to carry cargo. The power unit is a nonarticulated truck with one or more drive axles that carries either a detachable or a permanently attached cargo box. The trailing unit is a trailer or semitrailer with a cargo box so designed that the power unit may be loaded and unloaded through the trailing unit.

集装箱车辆：一种组合车，包括牵引车和挂车。它们的设计都用来装载货物。牵引车是一种不分节的车体，它有一个或多个车轴，既能装运可分离的货箱也能装运不可分离的货箱。挂车是一个含有集装箱的拖车或半拖车。这样设计便于牵引车通过挂车装卸货物。

Maximum Extent Practicable: The limits of available technology and the practical and technical limits on a pipeline operator in planning the response resources required to provide the on-water recovery capability and the shoreline protection and cleanup capability to conduct response activities for a worst case discharge from a pipeline in adverse weather.

最大可行程度：在可用技术和实际操作的限制下，管道经营公司在规划管道运作系统的应变能力时，要考虑到水上可恢复的能力和海岸线保护，还有在最坏情况下，恶劣天气中管道溢露事件的处理。

Means of Transportation: A mode used for going from one place (origin) to another (destination). Includes private and public modes, as well as walking. For all travel day trips, each change of mode constitutes a separate trip.

出行方式：从出发地到目的地的所用的交通方式。包括私人交通、公共交通和步行。对于所有的日出行，每次出行方式的变化将组成一个独立的出行。

English	Chinese
Median Category: Inclusion of a median within single instance of road.	**中央分隔带分类**：某一路段的中央分隔带的归属划分。
Median Included: Median is included within the instance of road.	**包含中央分隔带**：某一路段带有中央分隔带。
Median Not Included: Median is not included because there is no median or median is wide enough to cause separate instances of road.	**不包含中央分隔带**：中央分隔带没有被包含在路段中。原因是没有中央分隔带或中央分隔带不够宽而无法将上下行道路分离。
Medium or Heavy Trucks: A motor vehicle with a Gross Vehicle Weight Rating (GVWR) greater than 10,000 pounds (buses, motor-homes, and farm and construction equip-ment other than trucks are excluded).	**中等或重型卡车**：车辆总重（GVWR）大于一万磅（公共汽车，房车和农业及施工设备）。
Medium Speed: A speed not exceeding 40 miles per hour.	**中速**：不超过 40 英里/时 的速度
Megawatt: See also Electricity, Gigawatt, Kilowatt.	**兆瓦**：参见电学，亿瓦，千瓦
Megawatt Electric (MWE): One million watts of electric capacity.	**兆瓦电流 (MWE)**：一百万瓦特的电容量。
Metered Data: End-use data obtained through the direct measurement of the total energy consumed for specific uses within the individual household. Individual appliances can be submetered by connecting the recording meters directly to individual appliances.	**仪表测量数据**：最终使用数据，通过直接测量单个家庭为特定用途所消耗的能源得到。可以通过直接连接测量仪器与个人用具来实现辅助测量。
Metric: Refers to the modernized metric	**米制**：现代的米制单位体系，

English	Chinese

system known as the International System.

通常称为国际单位体系。

Metric Ton: A unit of weight equal to 2,204.6 pounds.

公吨：每个单位重为 2204.6 磅。

Metropolitan Planning Area: The geographic area in which the metropolitan transportation planning process required by 23 U.S.C. 134 and section 8 of the Federal Transit Act (49 U.S.C. app. 1607) must be carried out.

大都市规划区：23 U.S.C. 134 及第八联邦法案所规定的城市运输规划程序必须在该地理区域内执行。

Metropolitan Planning Organization (MPO): The forum for cooperative transportation decisionmaking for a metropolitan planning area.

大都市规划组织：为某一主要城市规划区制定联合交通决策的机构。

Metropolitan Planning Organization (MPO): Formed in cooperation with the state, develops transportation plans and programs for the metropolitan area. For each urbanized area, a Metropolitan Planning Organization (MPO) must be designated by agreement between the Governor and local units of government representing 75% of the affected population (in the metropolitan area), including the central cities or cities as defined by the Bureau of the Census, or in accordance with procedures established by applicable State or local law (23 U.S.C. 134(b)(1)/Federal Transit Act of 1991 Sec. 8(b)(1)).

大都市规划组织：与州政府协作，为主要城市规划区域制定交通规划和项目。对于每个市区，城市规划组织必须由州长和能够代表 75% 受影响人口的当地政府共同指定。受影响地区包括中心城市，或由人口普查局所规定的城市，或符合适用的国家或当地法律程序所建立的城市。

Metropolitan Statistical Area (MSA): Areas defined by the U.S. Office of Management and Budget. A Metropolitan Statistical Area

大都市统计地区：由美国管理预算局所定义的地区。大都市统计地区是 1）一个郡或一组

English	Chinese
(MSA) is 1) A county or a group of contiguous counties that contain at least one city of 50,000 inhabitants or more, or 2) An urbanized area of at least 50,000 inhabitants and a total MSA population of at least 100,000 (75,000 in New England). The contiguous counties are included in an MSA if, according to certain criteria, they are essentially metropolitan in character and are socially and economically integrated with the central city. In New England, MSAs consist of towns and cities rather than counties.	相邻的包括至少一个有五万居民或更多人口的郡。2）包括至少五万居民的城市地区，或者至少有十万人口的主要城市统计地区。如果按照某种标准，相邻的郡都被包括在大都市统计地区之内。这些郡基本上都是主要城市性质的，且与中央城市在社会和经济上相结合。在新英格兰地，大都市统计地区包括城镇和城市，而不是郡。
Metropolitan Status: A building classification referring to the location of the building either located within a Metropolitan Statistical Area (MSA) or outside a MSA.	**大都市状况**：建筑物分类，按照建筑物的位置，可能在大都市统计地区内部，也可能在外部，来确定。
Mexican Overdrive: Kicking out of gear going down grade.	**墨西哥超速传动**：下坡时离合器脱离而失控。
Midgrade Unleaded Gasoline: Gasoline having an antiknock index (R+M/2) greater than or equal to 88, or less than or equal to 90, and containing not more than 0.05 grams of lead or 0.005 grams of phosphorus per gallon.	**中级无铅汽油**：汽油的抗爆系数大于或等于 88，少于等于 90，每加仑最多包含 0.05 克铅或 0.005 克磷。
Mile: A statute mile (5,280 feet). All mileage computations are based on statute miles.	**英里**：法定英里（5280 英尺）。所有英里数的计算都基于法定英里。
Mile Marker: A point on a feature indicating the distance, in miles, measured along the course or path of the feature from an established origin point on the feature.	**英里标记**：从一个设定的起源点沿着给定路径进行测量，以英里标记距离的特征点。

English	Chinese
Miles Per Gallon (MPG): A measure of vehicle fuel efficiency. Miles Per Gallon (MPG) represents "Fleet Miles per Gallon". For each subgroup or "table cell", MPG is computed as the ratio of the total number of miles traveled by all vehicles in the subgroup to the total number of gallons consumed. MPGs are assigned to each vehicle using the Environmental Protection Agency (EPA) certification files and adjusted for on-road driving.	英里每加仑：一种车辆燃料效率的度量标准。英里每加仑代表"每加仑驶过的距离"。对于每个分单元或"表格单元"，计算英里每加仑就相当于，在分单元里所有车辆行驶的所有英里数的速度除以所有消耗的加仑数量。采用环保所的证明文件来确定每辆车的英里每加仑，并依此来调整实际驾驶。
Miles Per Gallon (MPG) Shortfall: The difference between actual on-road Miles Per Gallon (MPG) and Environmental Protection Agency (EPA) laboratory test MPG. Miles Per Gallon (MPG) shortfall is expressed as gallons per mile ratio (GPMR).	英里每加仑(MPG) 短缺：实际每加仑行使距离和美国环保所的实验测试每加仑行驶距离之间的差异。每加仑短缺量用加仑/英里来表示。(GPMR).
Mill Capital: Cost for transportation and equipping a plant for processing ore or other feed materials.	工厂资金：交通运输和装备矿石加工厂及其它供应材料的花费。
Mini Service: Service station attendants pump vehicle fuel but do not provide other services, such as checking oil and tire pressure or washing windshields.	基本服务：服务员只给车加油，不提供诸如检查油压，气压，和洗挡风玻璃等服务。
Minibridge: A joint water, rail or truck container move on a single Bill of Lading for a through route from a foreign port to a U.S. port destination through an intermediate U.S. port or the reverse.	短桥运输：一种水运、铁路和货物集装箱运输的联合运输方式，并且整个行程（包括从一个外国港口出发，通过美国中转站港口，到达美国目的地港口，或相反的过程）只使用一个提货单。

English

Minimum Descent Altitude: The lowest altitude, expressed in feet above mean sea level, to which descent is authorized on final approach or during circle-to-land maneuvering in execution of a standard instrument approach procedure, where no electronic glide slope is provided.

Minivan: New type of small van that first appeared with that designation in 1984. Any of the smaller vans built on an automobile-type frame. Earlier models such as the Volkswagen van are now included in this category.

Minor Alteration: An alteration other than a major alteration.

Minor Arterial: Streets and highways linking cities and larger towns in rural areas in distributing trips to small geographic areas in urban areas (not penetrating identifiable neighborhoods).

Minor Repair: A repair other than a major repair.

Miscellaneous Transport Revenue: Other revenues associated with air transportation performed by air carriers, such as transportation fees collected from those traveling on free or reduced transportation and processing service charges such as lost tickets.

Mixed Cargo: Indicates that a vessel carries any combination of grains, government aid,

Chinese

最低下降高度：最低高度，用高于平均海平面的高度表示。下降是指经过授权，或在未提供电子下滑坡度情况下按照标准方式执行绕场着陆。

小型面包车：新型面包车，在1984年首次命名。任何一种以机动车框架为基础而制造的小型面包车。早期的大众面包车就属于这一类。

小改造：除大改造之外的改造。

辅助干线：在农村地区，连接城市和大村镇的市政道路和公路，遍布城市中各个小范围的地理区域。（不穿过明确的居住小区）。

次修理：除大修外的修理。

混合交通收入：和空运相关联的其他收入，如从减免费用的运输所收取的交通费用，以及补票之类的手续费。

混装货船：指同时装运不同货物的船。货物种类包括：谷物、

English	Chinese

containers, general or bulk cargoes.

政府援助货物、集装箱、普通货物或散装货物。

Mobile Home: A housing unit built on a movable chassis and moved to the site. It may be placed on a permanent or temporary foundation and may contain one room or more. If rooms are added to the structure, it is considered a single-family housing unit. A manufactured house assembled on site is a single-family housing unit, not a mobile home.

活动房屋：建造在可移动底盘上并运抵目的地的房屋单元。它可以放置在永久或临时的基地上，可能包含一间或更多的房间。如果房间被添加到建筑物中，那么它就是一个单一的家庭房屋单元。现场装配的房屋是一个单一的家庭房屋单元，而不是活动房屋。

Mobile Home Park: An area maintained for the parking of inhabited mobile homes.

活动房屋区：一个可以持续停放有人居住的活动房屋的地区。

Modal Share: The percentage of total freight moved by a particular type of transportation.

模式份额：由某种特定运输方式承运的货物占货物总量的比例。

Modal Split: 1) The proportion of total person trips that uses each of various specified modes of transportation. 2) The process of separating total person trips into the modes of travel used. 3) A term that describes how many people use alternative forms of transportation. It is frequently used to describe the percentage of people who use private automobiles, as opposed to the percentage who user public transportation.

模式分配：1）采用不同交通模式的个人出行占全部个人出行的比例。2）分配个人出行模式的过程。3）一个用于描述采用可选择交通方式的人数的词汇。常用来描述使用私家车的人口比例，该比例与使用公共交通的人口的比例相加为1。

Mode: Any of the following transportation methods; rail, highway, air, or water.

出行方式：下列交通方式的任何一种：铁路交通、公路交通、空运交通和水运交通。

English	Chinese
Mode: Transit service operated in a particular format. There are two types: fixed-route and non-fixed route.	**公交服务方式**：特定形式的运输服务。有两种类型：固定路径及非固定路径。
Mode: Transportation planners, analysts, and decisionmakers refer to the means of transportation as a mode.	52 **出行方式**：交通规划者、分析者、决策者称交通方式为方式。
Moped: Includes motorized bicycles equipped with a small engine, typically 2 horsepower or less. Minibikes, dirt bikes, and trail bikes are excluded. Note that a motorized bicycle may or may not be licensed for highway use.	**机动脚踏两用车**：包括装有小型发动机（一般为两马力或更少）的机动化自行车，不包括小型机车，泥地摩托车，拖挂自行车。机动化的自行车得到或没有得到高速公路上行驶的许可都是可能的。
Most Harmful Event: The event during a crash for a particular vehicle that is judged to have produced the greatest personal injury or property damage.	**最大伤害事件**：被判产生最大人身伤亡和财产损失的特殊的机动车碰撞事件。
Most Restrictive State: The mode of an electric or electronic device that is equivalent to a track relay in its deenergized position.	**最大限制状态**：等同于继电器断开位置的电子装置模式。
Motor Carrier: 1) A for-hire motor carrier or a private motor carrier of property. The term "motor carrier" includes a motor carrier's agents, officers and representatives as well as employees responsible for hiring, super-vising, training, assigning, or dispatching of drivers and employees concerned with the installation, inspection, and maintenance of motor vehicle equipment and/or accessories. 2) An employer firm that is primarily engaged in providing commercial motor freight or long	**汽车运输商**：1）出租货运车辆或私人货运车辆。汽车运输商包括车辆代理商、职员、代表及负责出租、监督、训练、分配或发送司机的雇员和与装备、检查和汽车及其零部件维护有关的雇员。2）提供商用货运、长途运输或转运业务的公司。

distance trucking or transfer services.

Motor-Driven Cycle: A motorcycle with a motor that produces 5 brake horsepower or less.

机动自行车：产生小于等于五马力动力的机动车。

Motor Freight Transportation Warehousing and Stockyards: Includes establishments that provide local or long-distance trucking or transfer services, warehousing and storage of farm products, furniture or other household goods, and commercial goods of a general nature. The operation of terminal facilities for handling freight, with or without maintenance facilities is also included. Stockyards, establishments that provide holding pens for livestock in transit, are included in this major group. These stock yards do not sell or auction livestock.

机动车货运仓库和牲畜围栏：在本地或长途货运或转运服务中，储存农产品、家具或其他家用商品、一般商业用品的设备，以及货运仓库搬运货物的设施，维护设备也包括在内。牲畜围栏也包括在其中，用于保护运输中的牲畜。但不销售货物或拍卖牲畜。

Motor Gasoline: A complex mixture of relatively volatile hydrocarbons, with or without small quantities of additives, obtained by blending appropriate refinery streams to form a fuel suitable for use in spark ignition engines. Motor gasoline includes both leaded and unleaded grades of finished motor gasoline, blending components, and gasohol.

动力汽油：适当馏分拌合后得到的挥发性烃的混合物，可能含有少量添加剂，适合发动机火花塞点火的燃料。动力汽油既包括有铅汽油又包括高等级无铅汽油、混合成分和酒精汽油。

Motor Home: Includes self-powered recreational vehicles (RV) that are operated as a unit without being towed by another vehicle (e.g., a Winnebago motor home).

居住式旅游车：不需其他车辆牵引，能够提供自主动力的休闲车辆。（例如：温尼贝戈人居住式旅游车）。

English

Motor Vehicle: A vehicle, machine, tractor, trailer, or semitrailer, or any combination thereof, propelled or drawn by mechanical power and used upon the highways in the transportation of passengers or property. It does not include a vehicle, locomotive, or car operated exclusively on a rail or rails, or a trolley bus operated by electric power derived from a fixed overhead wire, furnishing local passenger transportation similar to street-railway service.

Motor Vehicle: Any mechanically or electrically powered device not operated on rails, upon which or by which any person or property may be transported upon a land highway. The load on a motor vehicle or trailer attached to it is considered part of the vehicle.

Motor Vehicle Accident: An unstabilized situation that includes at least one harmful event (injury or property damage) involving a motor vehicle in transport (in motion, in readiness for motion or on a roadway, but not parked in a designated parking area) that does not result from discharge of a firearm or explosive device and does not directly result from a cataclysm.

Motor Vehicle Chassis: The basic operative motor vehicle, including engine, frame, and other essential structures and mechanical parts, but excluding body and all accessories and auxiliary equipment.

Chinese

机动车：车辆、机器、牵引车、挂车或半挂车或任何其中的联合，由机械力推进或拖拽并用于道路客运或货运。不包括在轨道上运行的车辆、机车或汽车，也不包括顶部安装电线由电力驱动的类似于城市轨道旅客运输服务的无轨电车。

机动车：不在钢轨上而在道路上运输人和货物的机械或电力装置。机动车的货物或拖车都被认为是车辆的一部分。

机动车事故：非静止状态下包括人员伤亡或财产损失之一的伤害性事件。事故由运行中的机动车辆（启动、准备启动或在途行驶，而不是停放在一指定的停车场区域）引起，不是有由武器或引爆装置引起的，也不是直接由灾害等引起的。

机动车底盘：机动车运行的基础，包括发动机、车架和其他主要的结构和机械零部件，但不包括车体和其他零部件和辅助设备。

English	Chinese

Motor Vehicle In Transport: A motor vehicle in motion on the traffic way or any other motor vehicle on the roadway, including stalled, disabled, or abandoned vehicles.

途中运输车辆：在道路上行驶的机动车或任何其他在道路上行使的车辆，包括熄火车辆、堵塞车辆和自由运输的车辆。

Motor Vehicle Traffic Accident: An accident involving a motor vehicle in use within the right-of-way or other boundaries of a trafficway open for the use of the public.

机动车交通事故：涉及道路或其他公共路段上任何使用中机动车的事故。

Motorcycle: All two or three wheeled motorized vehicles. Typical vehicles in this category have saddle type seats and are steered by handle bars rather than a wheel. This category includes motorcycles, motor scooters, mopeds, motor powered bicycles, and three wheeled motorcycles.

摩托车：所有两轮或三轮的机动车辆。这一类型的典型车辆具有骑乘型的座位，用把手来控制方向而不是用方向盘。这一类型车辆包括机动摩托车、小型摩托车、机动自行车、发动机推力脚踏车，和三轮摩托车。

Motorcycle: A two- or three-wheeled motor vehicle designed to transport one or two people, including motor scooters, minibikes, and mopeds.

摩托车：设计能够载一或两人的单轮或双轮机动车辆，包括轻型摩托车、轻型自行车和机动自行车。

Motorized Vehicle: Includes all vehicles that are licensed for highway driving. Specifically excluded are snow mobiles and minibikes.

机械化车辆：包括全部许可上路行驶的车辆，不包括机动雪橇和小型机车。

Movable Bridge Locking: The rail locks, bridge locks, bolt locks, circuit controllers, and electric locks used in providing interlocking protection at a movable bridge.

活动桥锁：在活动桥上提供联锁保护的钢轨锁、桥梁锁、螺钉锁、转换开关和电子锁。

Mu Locomotive: A multiple operated electric locomotive described in 49 CFR 229.4 paragraph (i)(2) or (3).

轻型牵引机：小功率牵引机被用于货栈拉两轴运输平板车和堆场牵引车。

English	Chinese
Multi Stop Body: Fully enclosed truck body with driver's compartment designed for quick, easy entrance and exit.	**全密封卡车车体**：能够快速进出的包含驾驶室的全密封卡车车体。
Multi-Trailer Five or Less Axle Truck: All vehicles with five or less axles consisting of three or more units, one of which is a tractor or straight truck power unit.	**小于等于五轴的多挂卡车**：由三个或多于三个车体组成的小于等于五个轴的车辆，其中之一是牵引车或普通卡车。
Multi-Trailer Seven or More Axle Truck: All vehicles with seven or more axles consisting of three or more units, one of which is a tractor or straight truck power.	**大于等于七轴的多挂卡车**：由三个或多于三个车体组成的大于等于七个轴的车辆，其中之一是牵引车或普通卡车。
Multi-Trailer Six Axle Truck: All six axle vehicles consisting of three or more units, one of which is a tractor or straight truck power-unit.	**六轴拖挂卡车**：由三个或多于三个车体组成的六轴车辆，其中之一是牵引车或普通卡车。
Multimodal Transportation: Often used as a synonym for intermodalism. Congress and others frequently use the term intermodalism in its broadest interpretation as a synonym for multimodal transportation. Most precisely, multimodal transportation covers all modes without necessarily including a holistic or integrated approach.	**多方式运输**：常用作联运的同义词。国会和其他一些团体经常使用联运方式作为多式联运的广义解释。多式联运覆盖所有的方式，不一定包括一个整体的或综合的方式。
Multipurpose Passenger Vehicle: A motor vehicle with motive power, except a trailer, designed to carry 10 persons or less which is constructed either on a truck chassis or with special features for occasional off-road operation.	**多功能小客车**：设计载客数小于等于 10 人（不包括牵引载客数）的机动车辆。建造在卡车底盘上或具有少量越野驾驶的性能。

English	Chinese

National Bridge Inventory (NBI): The aggregation of structure inventory and appraisal data collected to fulfill the requirements of the National Bridge Inspection Standards that each State shall prepare and maintain an inventory of all bridges subject to the National Bridge Inspection Standards.

国家桥梁目录（NBI）：收集的建筑物目录的集合和评估数据，以满足国家桥梁验收标准。每个州应该制定并且维护按照国家桥梁验收的所有标准检验桥梁的记录。

National Bridge Inspection Standards (NBIS): Federal regulations establishing requirements for inspection procedures, frequency of inspections, qualifications of personnel, inspection reports, and preparation and maintenance of a State bridge inventory.

国家桥梁验收标准（NBIS）：指联邦对验收程序、验收的频率、人员的资格、验收报告的要求制定的规则，以及国家桥梁目录的制定和保持。

National Cooperative Highway Research Program (NCHRP): The cooperative research, development, and technology transfer (RD&T) program directed toward solving problems of national or regional significance identified by States and the FHWA, and administered by the Transportation Research Board, National Academy of Sciences.

国家协同高速公路研究计划（NCHRP）：相关部门合作参与研究、发展和进行技术转化（RD&T）的计划，以解决由州和联邦公路局确认的具有国家和地区意义的重大问题。由国家科学院下的交通研究委员会统筹管理。

National Cooperative Transit Research and Development Program: A program established under Section 6a) of the Urban Mass Transportation Act of 1964, as amended, to provide a mechanism by which the principal client groups of the Urban Mass Transportation Administration can join cooperatively in an attempt to resolve near-

国家协同公共交通研究和发展计划：在 1964 年的城市人口交通法第 6a) 章的基础上制定的计划，以提供一种机制，使得（联邦）城市公交管理署可以参与到解决近期城市公交问题应用研究、发展、测试和评估中来。该计划由交通研究委

English	Chinese

term public transportation problems through applied research, development, testing, and evaluation. NCTRP is administered by the Transportation Research Board.

员会管理。

National Highway System (NHS): This system of highways designated and approved in accordance with the provisions of 23 U.S.C. 103b).

国家高速公路系统（NHS）：指根据 23 U.S.C.103b)条款指定并批准的高速公路系统。

National Highway Traffic Safety Administration (NHTSA): The Administration was established by the Highway Safety Act of 1970 (23 U.S.C. 401 note). The Administration was established to carry out a congressional mandate to reduce the mounting number of deaths, injuries, and economic losses resulting from motor vehicle crashes on the Nation's highways and to provide motor vehicle damage susceptibility and ease of repair information, motor vehicle inspection demonstrations and protection of purchasers of motor vehicles having altered odometers, and to provide average standards for greater vehicle mileage per gallon of fuel for vehicles under 10,000 pounds (gross vehicle weight).

国家高速公路交通安全部门：这个部门是依照 1970 年的高速公路安全法设立的。该部门的主要任务是执行国会的指示，减少由于在国家高速公路上发生车辆碰撞而造成的死亡、受伤和经济损失的总数量，而且提供车辆易损性和维修信息、车辆检查示范，并防止更改车辆的里程表以保护购买者。它还提供车辆总重一万磅以下的车辆每加仑燃油所能行驶平均英里数指标 。

National Income: The aggregate earnings of labor and property which arise in the current production of goods and services by the nation's economy.

国民收入：指当前国家经济在商品生产和服务中出现的人力和财产的总收入。

National Transportation System: An intermodal system consisting of all forms of

国家运输系统：联合运输系统，包括所有的运输方式，以

English	Chinese

transportation in a unified, interconnected manner to reduce energy consumption and air pollution while promoting economic development and supporting the Nation's preeminent position in international commerce. The NTS includes the National Highway System (NHS), public transportation and access to ports and airports.

统一的、相互联系的方式减少能源消耗和空气污染，并促进经济发展，维持国家在国际商业方面的领导地位。国家运输系统包括国家高速公路系统（NHS）、公共交通以及与港口和飞机场的连接设施。

Nation's Freight Bill: The amount spent annually on freight transportation by the nation's shippers; also represents the total revenue of all carriers operating in the nation.

国家货运开销：全国货运方每年在货物运输上花费的总费用，也反映了国家货运业的年度营业收入。

Nationwide Personal Transportation Survey (NPTS): A periodic national survey that provides comprehensive information on travel by the U.S. population, along with related socioeconomic characteristics of the tripmaker. The Nationwide Personal Transportation Survey (NPTS) is designed to allow an analysis of travel by characteristics of the trip (e.g., length, purpose, mode), the tripmaker (e.g., age, sex, household income) and the vehicle used (e.g., model year, vehicle type, make and model). NPTS surveys were conducted in 1969, 1977 and 1983 by the Bureau of Census (BOC) for the Department of Transportation (DOT). The 1990 NPTS was sponsored by a group of DOT agencies, specifically the Federal Highway Administration (FHWA), Federal Railroad Administration (FRA), National

全国性的个人交通调查（NPTS）：定期的全国范围调查，提供美国居民出行的全面信息，以及出行者相关的社会经济状况。此项调查是为分析出行的特性(比如：距离、目的、方式)、出行者的特性(比如：年龄、性别、家庭收入)以及所使用的车辆的特性(比如：制造年份，车辆类型、厂家和型号)而设计的。1969 年、1977 年和 1983 年人口普查局（BOC）为交通部（DOT）进行了全国个人运输调查。1990年的调查由交通部的下属机构资助，特别是联邦公路局（FHWA）、联邦铁路局（FRA）、国家高速公路交通安全局（NHTSA）、部长办公

English	Chinese
Highway Traffic Safety Administration (NHTSA), Office of the Secretary (OST), and the Federal Transit Administration (FTA). The survey was conducted for DOT by Research Triangle Institute. Information was collected on all trips taken by each household member age 5 and older during a designated 24-hour period, known as a "travel day," and on trips of 75 miles or more taken during the preceding 14-day period, known as the "travel period." The trip information was expanded to annual estimates of trips and travel. The survey encompassed trips on all modes of transportation for all trip purposes and all lengths.	室（OST）以及联邦公共交通局（FTA）。这次调查由三角研究所为交通部完成，调查收集了所有的每个五岁以上的家庭成员在指定 24 小时期间（称为"出行日"）所进行的出行，以及以前的 14 天期间（称为"出行期"）所发生进行的 75 英里或更远距离的出行的资料。将出行资料进行扩展，对一年的短途出行和长途出行进行预测。调查包括了所有出行目的、出行距离以及所有的交通方式。
Net Horsepower: The usable power output of an engine "as installed". Net horsepower is the gross horsepower minus the horsepower used to drive the alternator, water pump, fan, etc., at a specified rpm.	**净功率**：指安装的发动机输出的可用功率。在指定的转速下，净功率是总功率减去为了驱动交流发电机、水泵、风扇等的功耗。
Net Income or Loss Before Income Taxes: The Operating Profit or (Loss) which is operating revenues less operating expenses less nonoperating income and expense produces the Net Income, but before "nonrecurring items."	**税前净收入（或损失）**：指营业利润（或损失）。净收入是营业收入减去营业费用，减去非营业性收入和费用，但是在"非重复性项目"之前的收入。
Net Maximum Dependable Capacity: The gross electrical output measured at the output terminals of the turbine generator(s) during the most restrictive seasonal conditions, less the station service load.	**最大净可靠容量**：测量出的涡轮发电机输出终端在最大限制季节条件下的总电量输出，减去服务站自身的能耗。

English	Chinese
Net Weight: (See also Gross Weight) Weight of the goods alone without any immediate wrappings, (e.g., the weight of the contents of a tin can without the weight of the can).	净重：（与总重相关）指物品的重量，不包括任何的包装（比如：易拉罐装的物品的自身重量而不包括易拉罐的重量）。
New Vehicle: A vehicle which is offered for sale or lease after manufacture without any prior use.	新型车辆：为销售或出租而提供的、制造后未经使用过的车辆。
New Vehicle Storage: A Fleet Management System (FMS) inventory status indicating vehicles that are placed in storage when first received and are awaiting assignment.	新车辆库存：车辆管理系统（FMS）的存货数据，表明了首次接受并准备派发的车辆。
Night: From 6:00 p.m. to 5:59 a.m.	夜间：从下午 6 点到早上 5:59。
Non-Bulk Packaging: A packaging which has 1) A maximum capacity of 450 L (119 gallons) or less as a receptacle for a liquid; 2) A maximum net mass of 400 kg (882 pounds) or less and a maximum capacity of 450 L (119 gallons) or less as a receptacle for a solid; or 3) A water capacity of 454 kg (1000 pounds) or less as a receptacle for a gas as defined in 49CFR173.115.	散货包装：指具有以下特征的包装：1）对于液体容器，最大的容积是 450 升（119 加仑）或者更少。2）对于固体容器，最大净重为 400 千克（882 磅）或更少，而且最大的容积为 450 升（119 加仑）或更少。3）对于气体容器（如 49CFR173.115 的定义），最大的装水能力为 454 千克（1000 磅）。
Non-Collision Accident: A motor vehicle accident which does not involve a collision. Non-collision accidents include jackknifes, overturns, fires, cargo shifts and spills, and incidents in which trucks run off the road.	非碰撞事故：指没有发生碰撞的汽车事故。非碰撞事故包括：车体积压和对折、倾覆、起火、货物散落和溢出，以及卡车脱离道路的事故。

English

Non-Earthen Shore: A structure built of stone, brick, concrete, or other building materials, that borders a body of water and that is not otherwise classified.

Non-Motorist: Any person who is not an occupant of a motor vehicle in transport and includes the following: 1) Pedestrians, 2) Pedalcyclists, 3) Occupants of parked motor vehicles, 4) Others such as joggers, skateboard riders, people riding on animals, and persons riding in animal-drawn conveyances.

Non-Motorist Location: The location of non motorists at time of impact. Intersection locations are coded only if non motorists were struck in the area formed by a junction of two or more traffic ways. Non-intersection location may include non-motorists struck on a junction of a driveway/alley access and a named traffic way. Non-motorists who are occupants of motor vehicles not in transport are coded with respect to the location of the vehicle.

Non-Occupant: Any person who is not an occupant of a motor vehicle (e.g., pedestrian or pedal cyclest), or who is an occupant of a motor vehicle which is not in transport.

Non-Regulated Trucking: A carrier which is exempt from economic regulation, e.g. exempt agricultural shipments and private trucking operations.

Chinese

非土质水岸：指由石头、砖、混泥土、或者其它的建筑材料建造的构筑物，与水接近，而且没有其他类型的构筑物。。

非乘车者：指不在驾驶中的汽车里的任何人，包括：1）行人，2）骑自行车者，3）已停车的汽车里的人，4）其他人，比如慢跑者，溜滑板的人，骑动物的人，以及乘由动物驾驶的交通工具的人。

非乘者地点：指事故发生时非乘汽车者的地点。只有当非乘汽车者在两条或更多条交通线路交叉所形成的地方发生事故时，地点才会被编码。非交叉地点包括非乘汽车者在车道口/小路通道和已命名的车道上被碰撞的地点。当非乘汽车者出于不运行的车辆中时，根据车辆的地点编码。

非车载者：指任何不在汽车内的人（比如行人或骑自行车的人），或指在停驶车辆中的乘坐者。

无管制的货物运输：指免除经济管制的货物运输，比如：部分农业运输和私人货物运输。

English	Chinese
Noncollision Crash: A class of crash in which the first harmful event does not involve a collision with a fixed object, non-fixed object, or a motor vehicle. This includes overturn, fire/explosion, falls from a vehicle, and injuries in a vehicle.	**非碰撞性事故**：事故的一种类型，指最先发生的有害事件不涉及与固定物体、非固定物体或汽车相撞的事故。包括翻转、起火/爆炸、从汽车上掉落、以及在汽车里受伤。
Nonfatal Accident: A motor vehicle traffic accident that results in one or more injuries, but no fatal injuries.	**非死亡事故**：指导致一个或更多个人受伤但并未导致死亡的汽车交通事故。
Nonfatal Alcohol Involvement Crash: Alcohol-related or alcohol-involved if police indicate on the police accident report that there is evidence of alcohol present. The code does not necessarily mean that a driver, passenger, or non-occupant was tested for alcohol.	**无致命的涉及酒精的碰撞**：如果警察在警方事故报告中提供有酒精出现的证据，则说明该事故是与酒精相关的或涉及酒精的。这个代号并不一定意味着驾驶者、乘员或非拥有者接受了酒精测试。
Nonfatal Injury: A nonfatal injury is any traffic accident injury other than a fatal injury.	**非死亡受伤**：非死亡受伤是指除死亡性受伤外的任何交通事故受伤。
Nonfatal Injury Accident: A nonfatal injury accident is a traffic accident that results in nonfatal injuries.	**非死亡受伤事故**：非死亡受伤事故是指导致非死亡性受伤的交通事故。
Nonfatal Injury Accident: Accident in which at least one person is injured, and no injury results in death.	**非死亡受伤事故**：指至少有一人受伤，但没有导致死亡的受伤的交通事故。
Nonfatal Injury Accident Rate: The nonfatal injury accident rate is the number of nonfatal injury accidents per 100 million vehicle miles of travel.	**非死亡受伤事故比率**：非死亡受伤事故比率是每一亿汽车行驶英里非死亡受伤事故的数量。

English

Nonfatal (Most Serious) Injured: Are nonfatally injured persons whose injury is classified as incapacitating (as defined in the "Manual On Classification of Motor Vehicle Traffic Accidents," American National Standards Institute (ANSI) D16.1-1989). States may receive information about these injuries on the accident report forms as incapacitating, incapacitating injury, incapacitated, disabled, carried from scene, severe injury, critical nonfatal, major injury, or other similar terms.

Nonfatally Injured Person: A nonfatally injured person is one who suffers a nonfatal injury in either a fatal or nonfatal injury traffic accident.

Nonfatally Injured Persons Rate: The nonfatally injured persons rate is the number of nonfatally injured persons per 100 million vehicle miles of travel.

Nonmetropolitan: Households not located within Metropolitan Statistical Areas (MSA) as defined by the U.S. Office of Management and Budget.

Nonoperating Income and Expense: Interest income and expense, unusual foreign exchange gains or losses, and capital gains or losses in disposition or property and equipment.

Nonprecision Approach Procedure: A standard instrument approach procedure in

Chinese

非死亡（最严重）受伤：非死亡性的受伤人员所受的伤归于丧失能力一类（如汽车交通事故分类指南中定义，美国国家标准协会（ANSI）D16.1-1989）。各州接到事故报道中的受伤信息采用如下名词：丧失能力、丧失能力的受伤、已丧失能力、残废、从现场运走、严重受伤、临界死亡受伤、重大伤、或其它类似的词。

严重受伤人员：严重受伤人员是在死亡性或非死亡性交通事故中遭受严重受伤的人。

严重受伤人员比率：严重受伤人员比率是每亿汽车行驶英里严重受伤人员的数量。

非都市区的：指不在美国管理预算办定义的都市统计区（MSA）内的住户。

非生产性的收入和支出：包括利息收入和支出，不经常的外币汇率变动导致的盈利或亏损，以及财产和设备处理带来的收入和开销。

非精确接近程序：标准的仪器接近程序，不提供电子滑动倾

English	Chinese
which no electronic glide slope is provided.	斜。
Nonpriority U.S. Mail: Mail bearing postage for surface transportation that goes by air on a space available basis at rates lower than those fixed for priority (i.e., air) mail.	**非优先美国邮件**：需要支付地面运输邮资的邮件，比确定为优先级的邮件邮资要低。在空运有剩余能力的情况下，也可航空运输。
Nonrecurring Items: Discontinued operations, extraordinary items and accounting changes income or loss.	**非重复性项目**：指非持续的运营、特别的项目以及会计方法改变带来的收入或开支。
Nonresident Commercial Driver's License: A commercial driver's license (CDL) issued by a State to an individual domiciled in a foreign country.	**非国民商业驾照**：指州政府向定居国外的个人颁发的商业驾照。
Not Paved Surface: All surfaces other than asphalt or concrete.	**非铺砌表面**：指除了沥青或混凝土以外的所有土地表面。

English	Chinese

Oakie Blower: Air scoop on air intake to increase power.

涡流器：在空气入口用以增加动力的空气扰流装置。

Objects Not Fixed: Objects that are movable or moving but are not motor vehicles. Includes pedestrians, pedalcyclists, animals, or trains (e.g., spilled cargo in roadway).

可移动物体：除内燃机车以外的可移动物体，包括：行人、脚踏车、动物或货运列车。

Occupancy: The number of persons, including driver and passenger(s) in a vehicle. Nationwide Personal Transportation Survey (NPTS) occupancy rates are generally calculated as person miles divided by vehicle miles.

上座率：在车内的人数，包括司机和乘客。全国个人交通调查的上座率是用所有乘客乘坐的英里数除以车辆行驶英里数估计算得。

Occupant: Any person who is in or upon a motor vehicle in transport. Includes the driver, passengers, and persons riding on the exterior of a motor vehicle (e.g., a skateboard rider who is set in motion by holding onto a vehicle).

乘客：位于移动中的机动车辆内部或者上面的任何人，包括司机、乘客，也包括位于机动车辆外部载体上的人员（例如通过将滑板连接到机动车辆上而处于运动状态的个人）。

Occupational Illness: Any abnormal condition or disorder of a railroad employee, other than one resulting from injury, caused by environmental factors associated with his or her railroad employment, including, but not limited to, acute or chronic illnesses or diseases which may be caused by inhalation, absorption, ingestion or direct contact.

职业病：由于工作环境因素而非外伤引起的任何铁路职工的反常状态或身心机能的失调，其中包括但不局限于急性或慢性疾病或者由于吸入，食入或直接接触有毒气体或液体而引起的疾病。

Occupied Caboose: A rail car being used to transport nonpassenger personnel.

乘务车厢：被用于运输除乘客之外的火车工作人员的车厢。

Off-Road Vehicular Area: An area for the testing of, or use by, vehicles that are designed to travel across the terrain.

越野车区：测试专门用于跨越这种地形的车辆或者仅供这些车辆使用的区域。

English	Chinese
Office of the Secretary of Transportation (OST): The Department of Transportation is administered by the Secretary of Transportation, who is the principal adviser to the President in all matters relating to Federal transportation programs. The Secretary is assisted in the administration of the Department by a Deputy Secretary of Transportation, a Associate Deputy Secretary, the Assistant Secretaries, a General Counsel, the Inspector General, and several Directors and Chairmen.	**交通部长办公室**：美国联邦交通部门是由交通部长来管理的，他是总统关于联邦交通运输方案中最主要的顾问。交通部长的助手包括一个副部长，一名副部长助理、一名部长助理，法律顾问、总检查长、及几位部门主任和主管。
On-Road Mile Per Gallon (MPG): A composite miles per gallon (MPG) that was adjusted to account for the difference between the test value and the fuel efficiency actually obtained on the road.	**每加仑可行里程**：每加仑可行里程：由于燃油效率的实验值与道路实际测量值的不同，通过调整得到的综合每加仑可行英里数。
On-Time Performance: The proportion of the time that a transit system adheres to its published schedule times within state tolerances.	**正点运行**：公交系统在指定的误差范围内遵照其公布的时刻表运营的比例。
Open-Access Transportation: The contract carriage delivery of nonsystem supply gas on a nondiscriminatory basis for a fee generally subject to transportation tariffs which are usually on an interruptible service basis on first-come, first-serve capacity usage.	**开放式运输**：非系统燃料供给的合同是建立在非歧视基础上。其价格是由可中断服务为基点，先来者先用，并受到交通关税的制约。
Open-Body Type Vehicle: A vehicle having no occupant compartment top or an occupant compartment top that can be	**敞蓬车**：没有顶棚，或者使用者可根据个人喜好安装或拆卸顶棚的车辆。

installed or removed by the user at his convenience.

Open Motorboat: Craft of open construction specifically built for operating with a motor, including boats canopied or fitted with temporary partial shelters.

敞篷摩托艇：马达露天布置或只有简易遮盖物的摩托艇。

Open to Public Travel Road: A road must be available, except during scheduled periods, extreme weather or other emergency conditions, and open to the general public for use by four-wheel, standard passenger cars without restrictive gates, prohibitive signs, or regulation other than restrictions based on size, weight, or class of registration. Toll plazas of public toll facilities are not considered restrictive gates.

公共道路：除非是处于预先安排的特定阶段，极端天气或其他紧急情况，公路必须是可用的，开放给四轮汽车、没有限制通行的关卡，只有基于大小、重量、或登记类别的限制。公共收费设备的收费区不属于限制性障碍。

Operating Assistance: Financial assistance for transit operations (not capital expenditures). Such aid may originate with federal, local, or state governments.

运营补贴：运营方面的补贴，但不包括固定资产投入。这种补贴多来自联邦、地方政府或州政府。

Operating Cost: 1) Fixed operating cost: in reference to passenger car operating cost, refers to those expenditures that are independent of the amount of use of the car, such as insurance costs, fees for license and registration, depreciation and finance charges; 2) Variable operating cost: in reference to passenger car operating cost, expenditures which are dependent on the amount of use of the car, such as the cost of gas and oil, tires, and other maintenance.

运营成本：1）固定运营成本：对于小汽车而言，包括汽车使用量以外的开支,如保险费用、执照费、注册、折旧及财务费用；2）可变运营成本：对于小汽车而言,开支数额取决于所使用汽车的程度，诸如天然气和石油的成本、轮胎、维修等。

English	Chinese
Operating Expenses: The costs of handling traffic, including both direct costs, (such as driver wages and fuel) and indirect costs (e.g. computer expenses and advertising) but excluding interest expense.	**运营开支**：处理交通流量的花销，即包括直接费用（司机工资，燃油费），也包括间接花销（如计算机购置花费和广告费），但不包括利息。
Operating Expenses: Expenses incurred in the performance of air transportation, based on overall operating revenues and overall operating expenses. Does not include nonoperating income and expenses, nonrecurring items or income taxes.	**运营开支**：航空运输中的开支。包括所有的运营收入和开支。不包括非运营性授予和开支，非重复性项目和收入税。
Operating Profit or Loss: Profit or loss from performance of air transportation, based on overall operating revenues and overall operating expenses. Does not include nonoperating income and expenses, nonrecurring items or income taxes.	**运营盈利或损失**：航空运输中的开支。包括所有的运营收入和开支。不包括非运营性收入和开支，非重复性项目和收入税。
Operating Property and Equipment: Owned assets including capital leases and leaseholds which are used and useful to the air carrier's central business activity, excluding those assets held for resale, or inoperative or redundant to the air carrier's current operations. These assets include loans and units of tangible property and equipment that are used in air transportation services and services incidental thereto.	**运营财产及设备**：航空公司拥有的，其商业活动不可缺少的资产和租赁合同。那些为转售而寄存，不工作的，或与现有运营而重叠的资产除外。这些资产包括用于航空交通服务的贷款及有形财产和设备。
Operational Road: Usable road and intended for use.	**可使用的公路**：可使用或是即将投入使用的公路。
Operator: See also Driver.	**操作者**：见司机。

English	Chinese
Operator: A person who controls the use of an aircraft, vessel, or vehicle.	**操作员**：操控飞机、航船或车辆的人。
Opposing Signals: Roadway signals which govern movements in opposite directions on the same track.	**反方向信号**：给同一车道上相反方向的车辆看的信号。
Opposing Train: A train, the movement of which is in a direction opposite to and toward another train on the same track.	**反方向列车**：在同一轨道上开往对面火车到来方向的火车。
Origin: Starting point of a trip.	**起点**：旅程的起始位置。
Origin: The country in which the cargo was loaded and/or the transit originated.	**起点**：货物装载地或运输起始地的所在国家。
Originated Carload: An originated carload is one which is loaded and begins its journey on a particular railroad.	**源地货物**：在某条铁路上被装载的货物，为该铁路的源地货物。
Other Freeways and Expressways: All urban principal arterials with limited control of access not on the Interstate system.	**其他高速公路**：所有不在州际高速系统上的其它具有准入控制的城市主要干线。
Other Nonoperating Income and Expenses, Net: Income from investor controlled companies and nontransport ventures, equipment and investments in securities of others, interest income, gains or losses resulting from nonroutine abnormal changes in the rates of foreign exchange, and other nonoperating items except capital gains or losses and interest expense.	**其他非运营收入、开支及纯利**：来自由公司所控的投资者和进行非运营性质投资活动和非运营性质投资活动的收入，设备和其他安全方面的投资，利息收入，由于外汇利率和其他非营运项目的非常规改变带来的收入和损失，但不包括资本收益或损失以及利息开支。
Other Single-Unit Truck: A motor vehicle consisting primarily of a single motorized	**其他单股卡车**：主要由独个动力装置组成的带有两个以上轮

English	Chinese
device with more than two axles or more than four tires.	轴或四个以上车轮的动力车辆。
Other Transport Related Revenue: Revenues from services such as in-flight sales, rentals and sales of services, supplies and parts.	**其他与运输相关的收入**：来自于途中零售、设备租赁和销售服务,以及零件供应等的收入。
Other Truck: All trucks other than pickups, (i.e., dump trucks, trailer trucks, etc.).	**其他卡车**：除轻型货车外的所有卡车（如：垃圾车，拖车等）。
Other Vehicle: Consists of the following types of vehicles: 1) Large limousines (more than four side doors or stretched chassis), 2) Three-wheel automobiles or automobile derivatives, 3) Van-based motor homes, 4) Light-truck-based motor homes (chassis mounted), 5) Large-truck-based motor homes, 6) ATVs (all terrain vehicle, including dune/swamp buggy) and ATC (all terrain cycle), 7) Snowmobiles, 8) Farm equipment other than trucks, 9) Construction equipment other than trucks (includes graders), 10) Other type vehicles (includes go-cart, fork lift, city streetsweeper).	**其他车辆**：由以下几种车辆组成：1）大型豪华轿车（多于四扇侧门或具有延长的底盘）；2）三轮汽车或汽车的衍生物；3）可用面包车驱动的机动屋；4）可用轻型卡车驱动的机动屋（装有底盘）；5）可用大型卡车驱动的机动屋；6）各种地形专用车（包括沙丘，沼泽）和特殊地形专用自行车；7）雪上汽车；8）除卡车外的农业用车；9）除卡车外的建筑用途车辆；10)其他类型汽车（包括手推车、叉车、城市道路清洗车）
Out Riggers: Device used for increasing width of trailers.	**悬臂支架**：用于增加拖车宽度的设备。
Outboard Designated Seating Position: A designated seating position where a longitudinal vertical plane tangent to the outboard side of the seat cushion is less than 12 inches from the innermost point on	**外侧指定座位位置**：指定座位位置在车辆中长纵端面至外侧座位小于十二英寸的地方，伸向点就在汽车内表面之间的高度设计 H 点和肩膀的参考点，

English	Chinese
the inside surface of the vehicle at a height between the design H-point and the shoulder reference point and longitudinally between the front and rear edges of the seat cushion.	以及前方和后方之间的纵向边缘席位垫。
Outer Continental Shelf: Offshore Federal domain.	**外大陆架**：近海岸的联邦属地。
Over-The-Road Bus: A bus characterized by an elevated passenger deck located over a baggage compartment.	**长途巴士**：一种乘客的座位高于行李车厢的汽车。
Over-The-Road Trip: An intercity movement by commercial motor vehicle.	**长途旅行**：城市间的商业性质的汽车运输。
Overall (Ton Miles, Load Factor, Available Capacity, Etc.): The total of passenger plus nonpassenger traffic, i.e., the sum of passenger and baggage, freight, express, U.S. mail, and foreign mail.	**总体**：（吨英里，载荷因子,容量等）所有的客运以及非客运交通，例如乘客数、行李、货物、快件、美国信件及外国邮件的总和。
Overall Vehicle Width: The nominal design dimension of the widest part of the vehicle, exclusive of signal lamps, marker lamps, outside rearview mirrors, flexible fender extensions. and mud flaps, determined with doors and windows closed and the wheels in the straight-ahead position.	**整体汽车宽度**：车辆设计时车身的最宽部分的宽度。不包括信号灯、标示灯、外后视镜.和挡泥板,由关闭的门窗和正向的车轮决定。
Overlook Access: (See also Public Use Class II Road) A road that provides access to a pull-off area, having definite entrance and exit points.	**休息区连接路**：连接具有出入口的休息区的道路。
Overpack: An enclosure that is used by a single consignor to provide protection or	**外包装**：供独立发货人用来保护或处理包裹的，或用来合并

English	Chinese
convenience in handling of a package or to consolidate two or more packages. Overpack does not include a freight container.	两个或两个以上的包裹的外部包装。不包括集装箱。

Owner/Operator: Independent trucker who drives the vehicle for himself or on lease to a company.

物主/经营者：为自己或是租赁公司开车的自由卡车驾驶员。

Owning Agency: A federal agency having accountability for government owned motor vehicles. This term applies when a federal agency has authority to take possession of, assign or reassign the vehicle regardless of which agency is using the vehicle.

拥有机构：管理政府拥有车辆的联邦机构，具有占用、分配、再分配车辆的权利，不管其他机构是否正在使用这些车辆。

English	Chinese
P & D's: Pick up and deliveries of freight.	**提取发送**：货物的提取和发送。
Package Freight: In the historical tables, designates cargo consisting of miscellaneous packages generally unidentifiable as to commodity and carried only on vessels licensed by the respective Authorities in Canada and the United States. This commodity classification is no longer applicable.	**零担货运**：在过去分类中，运输那些混装在一起无法辨认商品种类的货物运输形式称为零担货运，托运箱必须获得加拿大和美国相关部门的授权。目前这种货物分类已不再适用。
Packaging: A receptacle and any other components or materials necessary for the receptacle to perform its containment function in conformance with the minimum packing requirements of 49CFR, Chapter I, Subchapter C.	**包装物**：符合联邦规章代码第一章 C 节规定的最低包装标准要求的容器，以及任何保证该容器实现储存功能的成分或材料。
Packing Group: A grouping according to the degree of danger presented by hazardous materials. Packing Group I indicates great danger; Packing Group II, medium danger; Packing Group III, minor danger.	**包装危险等级**：根据危险材料的危险程度进行的分组。包装危险等级 I 指非常危险，等级 II 指中度危险，等级 III 指轻度危险。
Panel Body: Small, fully enclosed truck body often used for small package delivery.	**厢式车**：体积小的全封闭式的货车，常被用来运送小型的包裹。
Park: (See also Parkway) A place or area set aside for recreation or preservation of a cultural or natural resource.	**公园**：（参见公园道路）设置成休闲场所或者为文化/自然资源所预留的一个地点或地区。
Park and Ride: An access mode to transit in which patrons drive private automobiles or ride bicycles to a transit station, stop, or	**停车换乘**：一种交通通行方式，使用者驾驶私人汽车或骑自行车到公交中心、中转站或

English	Chinese
carpool/vanpool waiting area and park the vehicle in the area provided for the purpose. They then ride the transit system or take a car-or vanpool to their destinations.	合乘汽车/上下班交通合用等候区，将车辆停放在该区域，然后乘坐公交系统或合乘汽车去往目的地。
Park and Ride Lot: Designated parking area for automobile drivers who then board transit vehicles from these locations.	**换乘停车区**：指定的用于汽车驾驶员等候换乘公交车辆的停车场。
Parking Area: An area set aside for the parking of motor vehicles.	**停车区域**：为停放机动车辆而留出的一片区域。
Parking Brake: A mechanism designed to prevent the movement of a stationary motor vehicle.	**停车制动**：防止已静止的汽车滑动的一种机械装置。
Parking Brake System: A brake system used to hold a vehicle stationary.	**停车制动系统**：用于保持车辆静止的制动系统。
Parkway: (See also Park) A highway that has full or partial access control, is usually located within a park or a ribbon of park-like developments, and prohibits commercial vehicles. Buses are not considered commercial vehicles in this case.	**公园道路**：（参见公园）全部或部分控制的公路，通常位于公园或园林开发带，并且禁止商业车辆通行。但不限制公共汽车通行。
Participating Agency: A federal department or agency which transferred (consolidated) vehicles to the Interagency Fleet Management System (IFMS).	**参与机构**：调度车辆到跨部门车辆管理系统的联邦部门或机构。
Particulates: Carbon particles formed by partial oxidation and reduction of the hydrocarbon fuel. Also included are trace quantities of metal oxides and nitrides, originating from engine wear, component	**粉尘**：由部分氧化物和碳氢燃料还原形成的碳粒子。也包括由发动机磨损、元件退化和无机燃料添加剂产生的微量金属氧化物和氮化物。在交通运输

English	Chinese
degradation, and inorganic fuel additives. In the transportation sector, particulates are emitted mainly from diesel engines.	领域，粉尘的排放主要来自柴油发动机。
Passenger: With respect to vessels and for the purposes of 49 CFR 176 only means a person being carried on a vessel other than: the owner or his representative; the operator; a bona fide member of the crew engaged in the business of the vessel who has contributed no consideration for his carriage and who is paid for his services; or a guest who has not contributed any consideration directly or indirectly for his carriage.	**旅客**：根据美国联邦法规 49 章 176 条对出行目的规定，乘客指除了车主或他的代表、司机、从事该交通工具商业运营工作的人员、乘坐该运输工具过程中为他人服务并得到报酬的员工，或为其旅行不负担任何直接或间接报酬的旅客以外乘坐该运输工具的人。
Passenger: Any occupant of a motor vehicle who is not a driver.	**乘客**：任何除驾驶员外乘坐机动车的人。
Passenger Automobile: A passenger automobile is any automobile (other than an automobile capable of off-highway operation) manufactured primarily for use in the transportation of not more than 10 individuals.	**载客汽车**：承载不多于十名旅客的汽车（不同于在非高速公路运营的汽车）。
Passenger Car: A unit of rolling equipment intended to provide transportation for members of the general public, including self-propelled cars designed to carry baggage, mail, express and passengers.	**客车**：为普通大众提供运输服务的车辆设备，包括专门运载行李、邮件、快递和乘客的机动车辆。
Passenger Car: A motor vehicle with motive power, except a multipurpose passenger vehicle, motorcycle, or trailer, designed for carrying 10 persons or less.	**客车**：除多用途小客车、摩托车、或挂车外带有动力的机动车以外的运输乘客的机动车辆，设计载运十人或十人以下。

English	Chinese
Passenger Car: Any motor vehicle that is a convertible; 2-door sedan, hardtop, or coupe; a 4-door sedan or hardtop; a 3-or 5-door hatchback coupe; an automobile with pickup body; or station wagon.	**客车**：任何敞篷机动车，双门家用、硬顶（非敞篷）、或双人小轿车，四门家用或硬顶（非敞篷）汽车，三或五门的仓门式后背的小轿车；轻便客货两用车；或小旅行车。
Passenger Car: Any motor vehicle that is an automobile, auto-based pickup, large limousine, or three-wheel automobile or automobile derivative.	**客车**：任何机动车辆，如汽车、小型货车、大型高级轿车，或三轮汽车及其它类汽车设备。
Passenger Car: Motor vehicles used primarily for carrying passengers, including convertibles, sedans, and station wagons.	**客车**：主要为载运乘客的机动车辆，包括敞篷车、家用轿车和小型旅行车。
Passenger Car Equivalence: The representation of larger vehicles, such as buses, as equal to a quantity of automobiles (passenger cars) for use in level of service and capacity analyses.	**客车等价参数**：在道路设施服务水平和通车容量分析中，用来将大型车辆（如公共汽车）数量转化成产生相同交通影响的小型客车等价数量单位的参数。
Passenger-Carrying Volume: The sum of the front seat volume and, if any, rear seat volume, as defined in 40 CFR 600.315, in the case of automobiles to which that term applies. With respect to automobiles to which that term does not apply, "passenger-carrying volume" means the sum in cubic feet, rounded to the nearest 0.1 cubic feet, of the volume of a vehicle's front seat and seats to the rear of the front seat, as applicable, calculated in 49 CFR 523.2 with the head room, shoulder room, and leg room	**乘客载运体积**：对于符合美国联邦法规 40 章 600.315 条款规定的客车，乘客载运体积指汽车前座体积与后座体积（如果存在后座）之和。对于不适用于这一条款的客车而言，乘客载运体积指汽车前座体积和前座后边的座位所占的体积之和（以立方英尺为单位，并保留一位小数），根据美国联邦法规 49 章 523.2 条款规定的计算方式，其顶部空间、肩部

English	Chinese
dimensions determined in accordance with the procedures outlined in Society of Automotive Engineers (SAE) Recommended Practice J1100a, Motor Vehicle Dimensions (Report of Human Factors Engineering Committee, Society of Automotive Engineers, approved September 1973 and last revised September 1975).	空间和腿部空间尺寸都按照汽车工程师学会建议法案 J1100a 中机动车辆尺寸实施（人为因素工程委员会、汽车工程师学会编制的于 1973 年 9 月通过、1975 年 9 月校订完成的一份报告）所总结的规程来计算。
Passenger Count: A count of the passengers on a vehicle or who use a particular facility.	**乘客数量**：在汽车上的或使用某特定车辆设备的乘客数量。
Passenger Flow: The number of passengers who pass a given location in a specified direction during a given period.	**乘客流量**：在一定的时间段内按特定的方向通过某地点的乘客的数量。
Passenger-Miles Traveled (PMT): One person traveling the distance of one mile. Total passenger-miles traveled by all persons.	**乘客英里通行数**：一乘客英里通行数代表一个旅客通行一英里的基本单位。总乘客英里通行数用来计算所有乘客通行产生的乘客英里通行数。
Passenger Revenue: Money, including fares and transfer, zone, and park-and-ride parking charges, paid by transit passengers; also known as "farebox revenue."	**客运收入**：包括车费和转账、邮寄、换乘停车费等由乘客支付的金钱收入，亦称为"车票盒收入"。
Passenger Revenue Ton Mile: One ton of revenue passenger weight (including all baggage) transported one mile. The passenger weight standard for both domestic and international operations is 200 pounds.	**客运吨英里收入**：每吨英里的客运量（包括所有的包裹）收入。乘客标准限重在美国国内和国际上均规定是二百磅。
Passenger Service: Both intercity rail passenger service and commuter rail passenger service.	**客运服务**：包括城市间的轨道客运服务和通勤轨道客运服务。

English	Chinese
Passenger Vehicle: A vehicle with a Gross Vehicle Weight Rating (GVWR) of 10,000 pounds or less which includes passenger cars, light pickup trucks, light vans, and utility vehicles.	**客车**：车辆毛重量少于或等于一万磅的车，包括小轿车、小型载货卡车、轻型面包车和小型特殊用途车辆。
Passenger Vehicle Crash: A motor vehicle crash involving at least one passenger vehicle.	**客车事故**：涉及至少一辆客车的机动车事故。
Paved Road Surface: Bituminous, concrete, brick, block, and other special surfaces.	**铺筑路面**：沥青材料、混凝土、砖、石块和其他特殊材料铺筑的路面。
Paved Surface: Surface of asphalt or concrete.	**铺筑面**：由沥青或混凝土铺筑的表面。
Payload: The maximum load that a unit of equipment may carry within its total rated capacity. The payload is the Gross Vehicle Weight Rating (GVWR) less the tare weight or actual weight of the unloaded vehicle.	**有效负载量**：每单位设备在其总评估容量下的最大工作负载量。有效负载量是车辆总重减去自重或空载时的实际重量。
Payload: Weight of commodity being hauled. Includes packaging, pallets, banding, etc., but does not include the truck, truck body, etc.	**载重量**：托运商品的重量。包括包裹、货盘、包装带等等,但是不包括卡车和载重车身本身等重量。
Peak Kilowatt: One thousand peak watts.	**高峰千瓦**：一千个高峰瓦特。
Peak Megawatt: One million peak watts.	**高峰兆瓦**：一百万个高峰瓦特。
Peanut Wagon: Small tractor pulling a large trailer.	**花生货车**：小型拖车托大型挂车。

English	Chinese
Peat: Peat consists of partially decomposed plant debris. It is considered an early stage in the development of coal. Peat is distinguished from lignite by the presence of free cellulose and a high moisture content (exceeding 70 percent). The heat content of air-dried peat (about 50 percent moisture) is about 9 million Btu per ton. Most U.S. peat is used as a soil conditioner. The first U.S. electric power plant fueled by peat began operating in Maine in 1990.	**泥煤**：泥煤由半腐烂植物的残体组成，是生成煤的一个前期物质。泥煤因其无纤维素和高湿物质含量（超过70%）而不同于褐煤。风干泥煤的热含量（大约50%潮湿物质量）有大约9百万Btu每吨。大多数美国泥煤被用作土壤调节剂。美国第一家以泥煤为燃料的发电厂，于1990年在缅因州开始运营。
Pedalcycle: Vehicle propelled by human power and operated solely by pedals; excludes mopeds.	**脚踏自行车**：由人力驱动且完全由踏板控制推进的车辆；不包括机动自行车。
Pedalcyclist: A person on a vehicle that is powered solely by pedals.	**脚踏自行车司机**：驾驶脚踏自行车辆的人。
Pedestrian: Any person not in or upon a motor vehicle or other vehicle.	**行人**：未乘坐机动车或其他车辆的人。
Pedestrian Accident: Is any person injured or killed in a highway motor vehicle accident who is not a vehicle occupant.	**行人交通事故**：有行人在步行中受伤或死亡的机动车公路交通事故。
Pedestrian Fatality: Are the number of pedestrians fatally injured in motor vehicle accidents.	**行人死亡数**：在机动车事故中受致命伤害的行人数量。
Pedestrian Walkway (or Walkway): A continuous way designated for pedestrians and separated from the through lanes for motor vehicles by space or barrier.	**人行道（或步行道）**：一段连续的用于行人通过的专用道路，利用空间或者隔离障碍物与机动车的直行车道分离。

English	Chinese

Peg-Leg: Tandem tractor with only one power axle.

固定支柱拖拉机：仅有一个机动轮轴的串联拖拉机。

Pelvic Impact Area: That area of the door or body side panel adjacent to any outboard designated seating position which is bounded by horizontal planes 7 inches above and 4 inches below the seating reference point and vertical transverse planes 8 inches forward and 2 inches rearward of the seating reference point.

骨盘弹着区：车门或车身侧板临近指定座位外侧的区域，其水平面距参考点向上七英寸向下四英寸，垂直剖面距参考点向前八英寸向后二英寸。

Peninsula: A body of land jutting out into and nearly surrounded by water.

半岛：陆地的一部分延伸出水中几乎全部被水围绕的部分。

People Mover: An automated transportation system (e.g., continuous belt system or automated guideway transit) that provides short-haul collection distribution service, usually in a major activity center.

捷轨系统：自动交通系统（例如，传动带系统或自动导向交通系统）提供短距离的集散服务，通常设置在主要的活动中心。

Person: See also Pedestrian.

行人：参见行人。

Person Miles of Travel (PMT): A measure of person travel. When one person travels one mile, one person mile of travel results. Where 2 or more persons travel together in the same vehicle, each person makes the same number of person miles as the vehicle miles. Therefore, four persons traveling 5 miles in the same vehicle, make 4 times 5 or 20 person miles.

行人英里通行数：对行人交通的计量。当一个人移动一英里，即产生一行人英里数。当两个或者更多的人同时乘坐一辆车旅行，每个人产生与车辆英里通行数相同数量的行人英里通行数。所以，当四个人乘坐同一车辆旅行五英里时，生成斯乘以五或二十行人英里通行数。

Person Trip: A person trip is a trip by one or more persons in any mode of transportation.

行人出行：一人或多人采用任何交通方式产生的出行。每一

English	Chinese
Each person is considered as making one person trip. For example, four persons traveling together in one auto make four person trips.	个人被认为是产生一个行人出行。例如：四人同乘一辆汽车出行产生四个行人出行。
Personal Consumption Expenditure (PCE): As used in the national accounts, the market value of purchases of goods and services by individuals and nonprofit institutions and the value of food, clothing, housing, and financial services by them as income in kind. It includes the rental value of owner-occupied houses but excludes purchase of dwellings, which are classified as capital goods (investment).	**个人消费支出**：被用于国民核算的个人和非盈利机构进行货物和服务购买的市场价值，和他们视为收入的食品、服装、住房及金融服务的价值。它包括私人占有房产的租赁价值，但是不包括住所购买，因为这些住所被归类为商品（投资）资本。
Petroleum: Crude oil, condensate, gasoline, natural gasoline, natural gas liquids, and liquefied petroleum gas.	**石油**：原油、凝析油、汽油、天然汽油、天然气液和液化石油气。
Petroleum: A generic term applied to oil and oil products in all forms, such as crude oil, lease condensate, unfinished oils, petroleum products, natural gas plant liquids, and nonhydrocarbon compounds blended into finished petroleum products.	**石油**：所有油类和油类产品的通称，例如原油、矿场气凝析油、半成品油、石油产品、天然气处理装置液和非烃混合石油产品。
Pick-Up: Small delivery truck.	**小货车**：小吨位运货汽车。
Pickup Truck: Includes compact and full-size pickup trucks.	**小载货卡车**：包括小型的和中型的小载货卡车。
Pickup Truck: A motorized vehicle, privately owned and/or operated, with an enclosed cab that usually accommodates 2-3 passengers and an open cargo area in the rear.	**小载货卡车**：由私人拥有或营运的机动车辆，包括通常容纳二，三名乘客的单独客舱，并在车后有一个敞篷的货舱。小

English	Chinese
Pickup trucks usually have about the same wheelbase as a full-size station wagon.	载货卡车通常与中型小旅行车拥有一样的轴距。
Pig: Trailer transported on flat car.	**拖车**：装置在平板车上运输的拖车。
Piggyback: (See also Intermodal) A transportation concept whereby truck trailers are hauled on railroad flatcars.	**背负式运输**：（参见联运方式）载重拖车由铁路平板车托运的运输方式。
Piggyback: The transportation of highway trailers or removable trailer bodies on rail cars specifically equipped for the service. It is essentially a joint carrier movement in which the motor carrier forms a pickup and delivery operation to a rail terminal, as well as a delivery operation at the terminating rail head.	**背负式运输**：将公路用拖车或可装卸式拖车体安装在提供专门运输服务的铁路托运车上运输的货运方式。它实质上是一种公路车辆向铁路转运站集货和散货、以及在铁路终端转运点上处理配送业务的营运方式。
Piggyback Trailers: Trailers which are designed for quick loading on railcars.	**背负式联运拖车**：设计用于轨道列车进行快速装载的拖车。
Pigtail: Cable used to transmit electrical power to trailer.	**拖车电缆**：用于传送电功率到拖车上的电缆。
Pike: Turnpike.	**收费道路**：收费公路。
Pin-Up: Hook tractor to semitrailer.	**托钩**：将挂车钩在拖车上。
Placarded Car: A rail car which is placarded in accordance with the requirements of 49 CFR 172 except those cars displaying only the FUMIGATION placards as required by 49 CFR 172.510.	**布告汽车**：根据美国联邦法规49.172 的要求布告的机动有轨车，不包括按 49.172.510 美国联邦法规规定陈列的汽车。
Place: An area with definite or indefinite boundaries.	**地方**：有明确边界或无明确边界的地区。

English	Chinese
Plain: A region of generally uniform slope, comparatively level, and of considerable extent.	**平原**：具有相同坡度，比较平坦和相对宽广的区域。
Platform Body: Truck or trailer body with a floor, but no sides or roof.	**平板车身**：带有平板但没有侧栏或顶棚的卡车或拖车。
Play: Any free movement of components.	**窜动**：元件的任何自由运动。
Ply Rating: A measure of the strength of tires based on the strength of a single ply of designated construction. A 12 ply rating does not necessarily mean that 12 plies are present, only that the tire has the strength of 12 standard plies.	**轮胎层级**：用于衡量轮胎强度的指标。 此标准基于指定结构的单厚度轮胎力量的测量。轮胎层级为十二并不一定意味着轮胎由十二层建构，而只是意味着轮胎具有标准十二层级的强度。
Point Detector: A circuit controller which is part of the switch operating mechanism and operated by a rod connected to a switch, derail or movable point frog to indicate that the point is within a specified distance of the stock rail.	**点测器**：一种转换开关，属于传动机构开关的一部分，由连接到开关的杆来控制，脱轨器和活动的蛙点显示这个点在基本轨的特定距离之内。
Poison Gas In Bulk: The transportation, as cargo, of any poison gas in any quantity.	**集装毒气**：作为货物运输的任意数量的有毒气体。
Pole Trailer: A motor vehicle without motive power designed to be drawn by another motor vehicle and attached to the towing vehicle by means of a reach or pole, or by being boomed or otherwise secured to the towing vehicle, for transporting long or irregularly shaped loads such as poles, pipes, or structural members capable generally of sustaining themselves as beams between the supporting connections.	**长货挂车**：为运输长的或不规则的货物而设计的一种由机动车托拽和用绳或杆或其它安全的方式附于牵引车的没有动力的车辆，例如杆、管或在支撑物之间作为衡梁的构件。

English	Chinese
Pool Site: One or more spent fuel storage pools that has a single cask loading area. Each dry cask storage area is considered a separate site.	**储集区**：一个或多个废燃料储存区拥有一个桶形的装载区，每个干涸的桶形装载区被认为是一个单独的区域。
Post: A long, relatively slender and generally round piece of wood or other material.	**桩子**：长的、相对细的、通常是圆形的木质或其它材质的柱子。
Pots: Flares placed on highway to warn traffic of an obstruction or hazard.	**交通警报信号**：高速公路上的闪光警报，提示发生交通堵塞或出现危险情况。
Pour On the Coal: Step on the gas.	**加油**：踩油门。
Power Brake: Open throttle while applying brakes.	机动刹车：运用刹车时打开节流阀。
Power-Operated Switch: A switch operated by an electrically, hydraulically, or pneumatically driven switch-and-lock movement.	**机动开关**：由电力、水力或空气动力操纵控制开关运动的开关。
Power-Take-Off: A device usually mounted on the side of the transmission or transfer case, or off the front of the crankshaft, and used to transmit engine power to auxiliary equipment such as pumps, winches, etc.	**动力输出装置**：通常安装在传动器或变速箱上的装置，或脱离曲轴前端，用于传递发动机功率到辅助设备，如：水泵、曲柄等等。
Power Train: The group of components used to transmit engine power to the wheels. The power train includes the engine, clutch, transmission universal joints, drive shafts and rear axle gears.	**动力传动系**：传递发动机功率到车轮的一组元件，传动系包括发动机、离合器、万向节、传动轴和后轴齿轮。
Power Units: The control and pulling vehicle for trailers or semitrailers.	**动力单元**：对于挂车或半挂车控制和调节车辆的装置。

English	Chinese
Powered Axle: An axle equipped with a traction device.	**驱动轴**：带有牵引装置的轴。
Preferred Highway: See Preferred Route.	**首选公路**：参见首选线路。
Preferred Route: A highway for shipment of highway route controlled quantities of radio-active materials so designated by a State routing agency, and any Interstate System highway for which an alternative highway has not been designated by such State agency as provided by 49 CFR 177.826(b).	**首选线路**：由某政府路径机构和任何州际公路系统指定为运输大量的放射性物质时用的公路线路。 在没有按照由美国联邦法规提供的国家机关的指定认可的替代公路时，州际公路成为首选线路。
Premium Gasoline: (See also Gasoline) Gasoline having an antiknock index (R+M/2) greater than 90. Includes both leaded premium gasoline as well as unleaded premium gasoline.	**高级汽油**：（参见汽油）抗爆指数大于 90 的汽油、包括含铅高级汽油和优质无铅汽油。
Premium Grade Gasoline: (See also Gasoline) A grade of unleaded gasoline with a high octane rating, (approximately 92) designed to minimize preignition or engine "knocking" by slowing combustion rates.	**高等级汽油**：（参见汽油）高辛烷值（大约 92）甲级无铅汽油，通过减缓燃烧比率来降低自燃点或发动机爆震点。
Premium Leaded Gasoline: (See also Gasoline) Gasoline having an antiknock index (R+M/2) greater than 90 and contain-ing more than 0.05 grams of lead or 0.005 grams of phosphorus per gallon.	**高级含铅汽油**：（参见汽油）抗爆指数大于 90 的汽油，每一加仑含有超过 0.05 克的铅或 0.005 克的磷。
Premium Unleaded Gasoline: (See also Fuel, Gasohol, Gasoline, Kerosene) Gasoline having an antiknock index (R+M/2) greater than 90 and containing not more than 0.05 grams of lead or 0.005 grams of phosphorus per gallon.	**高级无铅汽油**：（参见燃料、乙醇的汽油溶液、汽油、火油）抗爆指数大于 90 的汽油，每一加仑含有不超过 0.05 克的铅或 0.005 克的磷。

English	Chinese

Preventive Maintenance (PM): (See also Maintenance, Maintenance Control Center) The systematic servicing and inspection of motor vehicles on a predetermined time, mileage or engine-hour basis. The period varies with the type of equipment and the purpose for which it is assigned.

预防性维修：（参见维修、维护控制中心）根据预定的时间、里程或发动机小时对机动车辆预先进行系统维护和检查，维护的周期根据机器型号和设计目的的不同而改变。

Primary Railway: Tracks providing a direct route through an area.

主要铁路：提供穿过某个区域的直接通路的铁路轨道。

Principal Arterial: (See also Arterial Highway, Minor Arterial) Major streets or highways, many with multi-lane or freeway design, serving high-volume traffic corridor movements that connect major generators of travel.

主干线：（参见干线公路、次级干线）主要街道或公路，大多是多车道或高速公路设施，服务于大运量的交通运输通道，连接主要的出行生成区域。

Principal Impact Point: The impact that is judged to have produced the greatest personal injury or property damage for a particular vehicle.

主要冲击点：评估某种车辆所产生最大的人身伤害或财产损失的冲击。

Principal Place of Business: A single location designated by the motor carrier, normally its headquarters, where records required by 49 CFR 387, 390, 391, 395, and 396 will be maintained. Provisions in this subchapter are made for maintaining certain records at locations other than the principal place of business.

主要营业地点：由运输机构指定的一个地点，通常是它的总部，在那里根据美国联邦法规49.387，390，391，395和396要求保存其确定的档案。此章节也为非主要营业地点的档案保管提供了规定。

Private Carrier: A commercial motor carrier whose highway transportation activities are incidental to, and in furtherance of, its primary business activity.

私人承运人：为基本商业活动偶然产生和推动的公路交通运输服务的商业机动车承运方。

English	Chinese
Private Entity: (See also Public Entity) Any entity other than a public entity.	**私人个体**：（参见公共个体）任何非公共的个体。
Private Fleet Vehicle: Ideally, a vehicle could be classified as a member of a fleet if it is: 1) Operated in mass by a corporation or institution, 2) Operated under unified control, or 3) Used for non-personal activities. However, the definition of a fleet is not consistent throughout the fleet industry. Some companies make a distinction between cars that were bought in bulk: rather than singularly, or whether they are operated in bulk, as well as the minimum number of vehicles that constitute a fleet (i.e. 4 or 10).	**私营汽车运输队**：理论上，被归为运输车队车辆的条件包括：1）由一家公司或部门在公共区域运营，2）统一控制管理，3）用于非个人活动。然而，运输队的定义在运输行业里的各个环节不是固定的。一些公司会根据车辆是否为成批购买获得来区分，或者根据车辆是否成批运营来区分，并且运输队车辆的最少数量也有规定（例如四辆或十辆）。
Private Motor Carrier (Of Passengers): A person who is engaged in an enterprise and provides transportation of passengers, by motor vehicle, that is within the scope of, and in the furtherance of that enterprise.	**私营（客运）运输商**：使用企业所有、为企业发展服务的机动车辆从事旅客运输企业经营的人。
Private Motor Carrier (Of Property): A person who provides transportation of property by motor vehicle, and is not a for-hire motor carrier.	**私营（财物）运输商**：使用机动车辆提供财物运输服务的人，一般不进行出租业务。
Private Road: Private road with restricted public use.	**私有道路**：限制公共使用的私有道路
Private Track or Siding: A track located outside of a carrier's right-of-way, yard, or terminals where the carrier does not own the rails, ties, roadbed, or right-of-way and includes track or portion of track which is devoted to the purpose of its user either by	**私有轨道或支线**：位于运输商路权、场地或站点之外的轨道，运输商不拥有钢轨、连接件和全部或部分轨道的路权，而以租赁方式或书面协议的方式使用。在这种情况下，租赁

English	Chinese
lease or written agreement, in which case the lease or written agreement is considered equivalent to ownership.	或书面协议被认为等同于所有权。
Private Transportation: 1) Any transport service that is restricted to certain people and is therefore not open to the public at large. 2) Owned or operated by an individual or group, not a governmental entity, for his or its own purposes or benefit.	**私有交通运输**：1）只限于特定的人使用而不向大众开放的运输服务。2）以个人或团体的私有利益为前提，由个人或团体拥有或经营的非政府机构。
Privately Owned Vehicle (POV): Employee's own vehicle used on official business for which the employee is reimbursed by the government on the basis of mileage.	**私有车辆**：雇员自有车辆用于公共事务，为此政府以行驶英里数向雇员进行报销。
Product: See also Cargo, Commodity, Freight, Goods.	**产品**：参见货物、商品、货运。
Project: A locally sponsored, coordinated, and administered program, or any part thereof, to plan, finance, construct, maintain, or improve an intermodal passenger terminal, which may incorporate civic or cultural activities where feasible in an architecturally or historically distinctive railroad passenger terminal.	**项目**：当地投资、协调并管理的项目，或其中的一部分，主要是规划、筹措资金、建造、维护或改善联运式客运枢纽站，从而在有建筑或历史特色的铁路客运站与市民生活或城市文化活动相协调。
Property Damage: The actual or estimated dollar value of vehicle, cargo, and other property damage incurred in motor vehicle accidents.	**财产损失**：在机动车辆事故中被破坏的车辆、货物以及其他财产的实际或估计的美元价值。
Property Damage Accident: An accident for which property damage of $4,400 or more, but no fatalities or injuries, was reported.	**财产损失事故**：官方记录的财产损失超过四千四百美元但无严重受伤或死亡的交通事故。

English	Chinese
Property-Damage-Only Crash: A police-reported crash involving a motor vehicle in transport on a trafficway in which no injuries of any severity, including fatal injuries, are reported.	**财产损失事故**：警方记录的没有任何严重受伤或死亡的机动车辆交通事故。
Property Damage Rate: The average amount of property damage per accident or per one hundred accidents.	**财产损失率**：每一起交通事故或每百起交通事故的平均财产损失数额。
Property Damage Threshold: The amount of property damage used to determine whether an accident not involving fatalities or injuries is reportable under the Federal Motor Carrier Safety Regulations (FMCSR). In 1994, the property damage threshold was $4,400.	**财产损失标准**：财产损失标准数额，用来决定无人身伤亡交通事故是否应该根据联邦机动车辆安全条例进行官方记录。1994 年的财产损失标准数额为四千四百美元。
Property Loss Prevented: Calculated estimate of the amount of property loss that would have occurred had the Coast Guard not rendered assistance. It is based upon value of property assisted in cases where severity of the incident was evaluated as severe or moderate in nature.	**财产损失挽回数**：所计算估计出来的在没有既定保护措施情况下可能产生的财产损失数额。这个数额是根据那些严重或中等严重程度的交通事故中所挽回的财产数额来计算的。
Provisional Rate-Density Relationship: The relationship between fatality rates and average daily traffic. It is based on data for the 4-year period preceding the calendar year for which detailed data are reported. It is labelled "provisional" to make it clear that it is to be used as a guide rather than a standard. A provisional rate-density relationship may be described graphically or mathematically by a rate-density curve.	**临时比率-密度关系**：死亡率和平均日交通量之间的关系。以四年一周期的数据为基础，优先于每年报告的详细数据。称为"临时"是指它只是作为指导而不是标准。临时密度比率关系可能以密度比率曲线的形式用图表或算术形式来描述。

English	Chinese
Public Authority: Means a Federal, State, county, town or township, Indian tribe, municipal or other local government or instrumentality thereof, with authority to finance, build, operate, or maintain highway facilities, either as toll or toll- free highway facilities.	**公共管理局**：指联邦政府、州政府、郡、镇或区、印第安部落、市政或其它地方政府或媒介，这些部门对当地融资、建设、运营或维护公路设施（无论是收费还是免费的公路设施）拥有管理权。
Public Crossing: A location open to public travel where railroad tracks intersect a roadway that is under the jurisdiction and maintenance of a public authority.	**公共交叉口**：铁路与公路交叉口，其公路由政府管辖和维护并公共交通开放。
Public Entity: (See also Private Entity) 1) Any state or local government; 2) Any department, agency, special purpose district, or other instrumentality of one or more state or local governments; and 3) The National Railroad Passenger Corporation (Amtrak) and any commuter authority.	**公共机构**：（参见私人个体）1）任何国家或地方政府；2）任何部门、机构、特殊用途区域或其他的一或多国家地方政府媒介；3）全国铁路客运公司和其他通勤机构。
Public Liability: Liability for bodily injury or property damage and includes liability for environmental restoration.	**公众责任**：对人身伤害或财产损失责任，以及环境的护理所需承担的责任。
Public Road: Any road under the jurisdiction of and maintained by a public authority and open to public travel.	**公用道路**：由当地政府维护，管辖范围并向公众开放的交通道路。
Public School Transportation: (See also School Bus) Transportation by school bus vehicles of school children, personnel, and equipment to and from a public elementary or secondary school and school-related activities.	**公共中小学运输**：（参见校车）由校车运送学生、员工和设备往返于公众场所、学校和学校相关机构的运输。

English	Chinese
Public Street and Highway Lighting: Includes electricity supplied and services rendered for the purpose of lighting streets, highways, parks, and other public places or for traffic or other signal system service, for municipalities or other divisions or agencies of state or Federal governments.	**街道和公路照明**：为街道、公路、公园和其他公共场所提供照明，或为交通、信号系统服务，为市区或国家政府部门机构服务提供照明的供电服务。
Public Transit: Passenger transportation services, usually local in scope, that is available to any person who pays a prescribed fare. It operates on established schedules along designated routes or lines with specific stops and is designed to move relatively large numbers of people at one time.	**公共交通**：通常在地方范围内，任何人付一定的费用即可享有的旅客运输服务。它按照定制的计划沿特定的线路运行，有特定的停靠站点，而且一次运输的乘客数量较多。
Public Use Class I Road: A principal road/rural parkway which constitute the main access route, circulatory tour, or thoroughfare for visitors.	**公用一级道路**：作为主要通行路线、循环路线或地方通行路线而运营的主要干道/乡村道路。
Public Use Class II Road: (See also Overlook Access) A connector road which provides access within an area of scenic, scientific, recreational or cultural interest, such as overlooks, campgrounds, etc.	**公用二级道路**：提供风景区、学术场所或文化娱乐场所（景区，营地等等）内部通路的连接道路。
Public Use Class III Road: A special purpose road which provides circulation within public use areas, such as camp-grounds, picnic areas, visitor center complexes, concessioner facilities, etc. These roads generally serve low-speed traffic and are often designed for one way circulation.	**公用三级道路**：在公共使用区域内提供交通路线的特殊用途道路，例如露营地、野餐区、游客综合中心、特许设施等。这些道路通常服务于低速行驶车辆而且往往是单向交通模式。

English	Chinese
Public Use Class IV Road: A primitive road.	**公用四级道路**：天然土路。
Public Use Road: All roads that are intended principally for the use of visitors for access into and within the public use area included. This includes all roads that provide vehicular passage for visitors, or access to such representative park areas as point of scenic or historic interest, campgrounds, picnic areas, lodge areas, etc. County, State, and U.S. numbered highways maintained by the National Park Service are included in this category for purposes of functional classification.	**公共使用道路**：主要为旅客交通出行服务的公共区域内的全部道路。包括所有为旅客提供机动车通行或通向例如风景名胜古迹、露营地、野餐区和寄宿区等休闲区域的道路。按照郡道、州道以及国道等级分类进行数字编码的由国家公园服务处维护的公路都属于这个范畴。
Public Way: Any public street, road, boulevard, alley, lane, or highway, including those portions of any public place that have been designated for use by pedestrians, bicycles, and motor vehicles.	**公众路线**：任何公共街道、道路、林荫道、小路、小巷或公路，包括那些公共场所行人、自行车和机动车辆所使用的道路。
Purchase or Lease: With respect to vehicles, means the time at which an entity is legally obligated to obtain the vehicles, such as the time of contract execution.	**购买或租约期**：就车辆而言，指法律上有权拥有车辆的时间，即按合同执行的时间。
Purchased Transportation: Transportation service purchased by a public agency from a public or private provider on the basis of a written contract.	**购获运输**：在书面契约的基础上，官方机构从公共或私人供应商那里购买的运输服务。
Put On the Air: Apply the brakes.	**刹车**：施行制动。
Put On the Iron: Put on tire chains.	**防滑**：添加防滑轮胎链。

Rag Top: Open top trailer covered with a tarpaulin.

Rags: Bad tires.

Rail-Highway Grade Crossing: (See also Grade Crossings; Highway-Rail Crossing) A location where one or more railroad tracks cross a public highway, road, or street or a private roadway, and includes sidewalks and pathways at or associated with the crossing.

Railbus: A relatively light, diesel-powered, two-axle rail vehicle with a body resembling that of a bus.

Ramp: An inclined roadway connecting roads of differing levels.

Ramp Metering: 1) The process of facilitating traffic flow on freeways by regulating the amount of traffic entering the freeway through the use of control devices on entrance ramps. 2) The procedure of equipping a freeway approach ramp with a metering device and traffic signal that allow the vehicles to enter the freeway at a predetermined rate.

Rating: A statement that, as a part of a certificate, sets forth special conditions, privileges, or limitations.

Ratio Estimate: (See also Estimate Ratio, Mean) The ratio of two population aggregates (totals). For example, "average

布顶：用防水油布盖着的顶部敞开的挂车。

抛锚：轮胎损坏。

铁路 - 公路平面交叉口：（又见平面交叉口;、高速公路交叉口）一条或多条铁路的轨道与公路、街道或私人道路相交的位置，包括在交叉口上或与交叉口相联的人行道和小径。

有轨公交车：一种相对较轻，柴油机驱动，带有像公共汽车车体的两轴轨道车辆。

匝道：联接不在同一平面的道路之间的倾斜路段。

匝道控制：1）通过在匝道入口处使用交通控制装置控制进入高速公路的交通流量来促进高速公路上交通流畅通的过程；2）在高速公路入口安装测量装置以及交通信号，从而使车辆以预先决定的频率进入高速公路的过程。

评级：作为证书的一部分，确定特殊条件、特权或限制。

比率估计：（又见估计比率，平均值）两个集合数之比。比如，"每辆车行驶的平均里程

English	Chinese
miles traveled per vehicle" is the ratio of total miles driven by all vehicles, over the total number of vehicles.	数"是所有车辆行使的里程数与车辆总数之比。
Rear Axle Capacity: The factor and/or Society of Automotive Engineers (SAE) recommended maximum load that a rear axle assembly is designed to carry as rated at the ground and expressed in pounds.	**后轮轴承载力**：美国汽车工程师学会推荐的后车轴在地面上的设计的最大装载重量，用磅表示。
Rear End Collision: 1) A collision in which one vehicle collides with the rear of another vehicle. 2) A collision in which the trains or locomotives involved are traveling in the same direction on the same track. 3) A collision of the front of one vehicle with the rear of another vehicle. Also called rear-end.	**尾部碰撞**：1）一辆汽车撞上另一辆汽车的尾部；2）在相同方向同一轨道行驶的火车或机车之间的碰撞；3）一辆汽车的前部与另一辆汽车的尾部之间的碰撞。又叫头尾相撞。
Rear Extremity: The rearmost point on a vehicle when the vehicle's cargo doors, tailgate or other permanent structure are positioned as they normally are when the vehicle is being driven. Non-structural protrusions such as tail lights, hinges, an latches are deleted from the determination of the rearmost point.	**尾部末端**：当汽车行驶时汽车的货物门，汽车挡板或其它永久结构物位于正常位置时车辆最后面的点。非结构突出物如尾灯、铰链、门插销在确定最后面的点时忽略不计。
Rear Overhang: Distance from the center of the rear axle to the end of frame.	**车尾悬空距**：后车轴中心到框架尾部的距离。
Rebuild: (See also Remanufactured Vehicle) A complete repair of a component with the objective of returning it as nearly as possible to its original and/or performance characteristics.	**重建**：（又见重造车辆）在尽可能恢复物体的原状和工作特性的目标下对物体的构件进行完全的修复。

English	Chinese
Receiver: A device on a locomotive, so placed that it is in position to be influenced inductively or actuated by an automatic train stop, train control or cab signal roadway element.	**接受器**：一种安装在机车上用于确定位置，能够被自动火车停车装置、火车控制器或驾驶室信号的道路单元感应影响或激活的装置。
Receiver Coil: Concentric layers of insulated wire wound around the core of a receiver of an automatic train stop, train control or cab signal device on a locomotive.	**接受器线圈**：机车上缠绕在自动停车装置、火车控制器或驾驶室信号装置铁芯上的同轴隔离层电线。
Receptacle: A containment vessel for receiving and holding materials, including any means of closing.	**容器**：用来接收或保存原料的控制器皿，包括各种封闭的器皿。
Reconciling Items: Items where accounting practices vary for handling these expenses as a result of local ordinances and conditions. Reconciling items include depreciation and amortization, interest payments, leases and rentals. They are called reconciling items because they are needed to provide an overall total that is consistent with local published reports.	**核对项目**：在项目中因为地方法律条令和条件的不同在处理这些支出时账目清算也不同。核对项目包括折旧、摊销、利息、租金、契约。称为核对项目是因为它们必须与当地的出版总报告一致。
Reduction: Used to indicate the slower output speed resulting from a ratio proportion (faster on reductions of less than 1); A. single reduction: a single set of reducing gears in the rear axle; B. double reduction: an additional gear-set in the rear axle to reduce output speed further. May or may not be used as a 2-speed rear axle.	**缩减量**：用于表明由于比率而产生的慢输出速度（小于1的缩减量较快）；A. 单减：后车轴上单套缩减装置；B. 双减：双套装置于后车轴以进一步减小输出速度。可能用作双速后车轴。
Refined Petroleum Products: Refined petroleum products include but are not	**石油提炼产品**：精制石油产品包括但不只限于汽油、煤油、

English	Chinese
limited to gasolines, kerosene, distillates (including No. 2 fuel oil), liquefied petroleum gas, asphalt, lubricating oils, diesel fuels, and residual fuels.	蒸馏物（包括二号燃油）、液化石油气、沥青、润滑油、柴油和残余燃料。
Reflective Material: (See also Left Bank, Reflex Reflector, Retro-Reflective Material, Right Bank) A material conforming to Federal Specification L-S300, "Sheeting and Tape, Reflective; Non-exposed Lens, Adhesive Backing," (September 7, 1965) meeting the performance standard in either Table 1 or Table 1A of Society of Automotive Engineers (SAE) Standard J594f, "Reflex Reflectors" (January, 1977).	**反光材料**：（又见左岸,反射性反光镜,复古反光材料,右岸)一种符合联邦规格 L - S300 的材料，"压片和胶带，反光的，非暴露镜头，背面粘合"，（1965 年 9 月 7 日）符合汽车工程师学会标准 J594f 表 1 或表 1 A 的性能标准。"反射性反光镜"（1977 年 1 月）
Reflex Reflector: (See also Reflective Material) A device which is used on a vehicle to give an indication to an approaching driver by reflected lighted from the lamps on the approaching vehicle.	**反射性反光镜**：（又见反光材料）汽车上使用的一种装置，通过反射驶近的汽车的灯光来指示靠近的司机。
Reformulated Motor Gasoline: (See also Gasoline) Motor gasoline, formulated for use in motor vehicles, the composition and properties of which are certified as reformulated motor gasoline by the Environmental Protection Agency.	**再生马达汽油**：又见"汽油"，马达汽油，专供机动车使用，其成分和性能被环境保护处认证为再生马达汽油。
Regional Distribution Port: Waterfront area which 1) Is identifiable with a Standard Metropolitan Statistical Area (SMSA) as defined by the U.S. Bureau of Census, 2) Has 10 or more commercial terminal facilities located within a reasonable distance of the	**区域性配送港口**：指那些符合如下条件之一的海港区 1）可以确定为由美国普查局定义的标准大都市统计区；2）在一定的合理范围内有十个以上商业终端设施；3）至少有二条

English	Chinese
general area, 3) Is served by at least two Class I railroads, and 4) Is served by at least five interstate or U.S. highways.	以上一级铁路参与运输 4) 至少有五条以上州际或国家公路参与运输。
Registered Inspector: A person registered with the Department [of Transportation (DOT)] in accordance with 49 CFR 107 Subpart F who has the knowledge and ability to determine if a cargo tank conforms with the applicable DOT specification and has, at a minimum, any of the combinations of education and work experience in cargo tank design, construction, inspection, or repair set out in 49 CFR 171.8.	**注册检查员**：按照 49 CFR 107 F 规定的在运输部注册过的人员，拥有知识和能力判断货箱是否符合运输部的适用标准，至少拥有 49 CFR 171.8 提出的在货箱设计、建造、检查、维修方面有教育和工作经验。
Regular Gasoline: See also Gasoline.	**普通汽油**：又见汽油。
Regular Grade Gasoline: (See also Gasoline) A grade of unleaded gasoline with a lower octane rating (approximately 87) than other grades. Octane boosters are added to gasoline to control engine preignition or "knocking" by slowing combustion rates.	**普通等级汽油**：（又见汽油）比其他等级汽油的辛烷值低（大约 87）的一种无铅级汽油。辛烷辅助物加在汽油中通过减缓燃烧速度以控制提前点火和爆震。
Regular Leaded Gasoline: (See also Gasoline) Gasoline having an antiknock index (R+M/2) greater than or equal to 87 and less than or equal to 90 and containing more than 0.05 grams of lead or 0.005 grams of phosphorus per gallon.	**普通含铅汽油**：（又见汽油）抗爆指数（R + M/2）大于等于 87 小于 90、每加仑汽油含有多于 0.05 克铅或 0.005 克磷的汽油。
Regular Unleaded Gasoline: (See also Gasoline) Gasoline having an antiknock index (R+M/2) greater than or equal to 85 and less than 88, and containing not more	**普通无铅汽油**：又见汽油，抗爆指数(R+M/2)大于等于 85 小于 88、每加仑汽油含有不多于 0.05 克铅或 0.005 克磷的汽

than 0.05 grams of lead or 0.005 grams of phosphorus per gallon.

油。

Regularly Employed Driver: A driver who, in any period of 7 consecutive days, is employed or used as a driver solely by a single motor carrier.

常规雇佣司机：在某时期至少连续 7 天被同一汽车运营公司单独雇佣或聘用的司机。

Regulated Motor Carrier: A carrier subject to economic regulation by the Interstate Commerce Commission.

受约汽车运输公司：受州际商务委员会制定的经济法规约束的运输公司。

Regulation: Any agency statement of general or particular applicability designed to implement, interpret, or prescribe policy in order to carry out the purpose of a law. Synonymous with "rule" it has the force of law.

规章：为了实现法律目标而实施、解释、制订政策而设计的具有一般和特殊适用性的任何机构的声明。

Relayed Cut-Section: A cut-section where the energy for one track circuit is supplied through front contacts or through front and polar contacts of the track relay for the adjoining track circuit.

传送截面：在这个截面上一个轨道电路的电能通过和毗邻轨道电路上的轨道继电器的前方接触点或前方和中心接触点提供。

Remanufactured Vehicle: A vehicle which has been structurally restored and has had new or rebuilt major components installed to extend its service life.

重建车辆：汽车在结构上重新修复，有了新的主要部件或重修了主要部件以延长它的服务期。

Renewable Energy: Energy obtained from sources that are essentially inexhaustible (unlike, for example, the fossil fuels, of which there is a finite supply). Renewable sources of energy include wood, waste, photovoltaic, and solar thermal energy.

可再生能源：能量获取的来源在本质上是无穷尽的（比如，矿物燃料不是可再生能源，因为它只能有限供应）。可再生能源包括木材、垃圾、光电能、太阳热能。

English	Chinese

Rental of Railroad Cars: Establishments primarily engaged in renting or leasing railroad cars, whether or not also performing services connected with the use thereof, or in performing services connected with the rental of railroad cars.

铁路车辆出租公司：主要从事有轨电车租赁的公司，不管是否提供使用有轨电车相关的服务，或者提供与有轨电车租赁相关的服务。

Replacement Standard: The estimated useful life of a motor vehicle expressed in time (months or years) and/or utilization (miles).

置换标准：估计的一辆汽车的大概使用寿命期，用时间（月或年）或/和使用率（英里）表示。

Replacement Vehicle: A vehicle acquired to replace a vehicle in inventory that meets the replacement standard or becomes uneconomical to retain in service.

置换车辆：用另一辆车取代一辆符合置换标准或继续使用已不经济的汽车。

Reportable Accident: A motor vehicle accident involving a carrier subject to the Department of Transportation Act, which results in a fatality, injury, or property damage of $4,400 or more.

可报告事故：根据交通部运输条例，造成伤亡或财产损失超过四千四百美元的汽车交通事故。

Reportable Damage: Includes labor costs and all other costs to repair or replace in kind damaged on-track equipment, signals, track, track structures or roadbed. Reportable damage does not include the cost of clearing a wreck; however, additional damage to the above listed items caused while clearing the wreck is to be included in your damage estimate. Examples of other costs included in reportable damage are: 1) Rental and/or operation of machinery such as cranes, bulldozers, etc., including the services of

可报告损失：包括劳动力成本，维修或更换损坏的轨道设备、信号设备、轨道、轨道构造或路基的的所有其它成本。可报告成本不包括对事故现场的清理成本。但是，当清理事故现场时，除上面列出的损失项目外的其他损失将被包括在你的损失估计中。其他损失包括在可报告损失中的例子有：1）租用或/和操作机器如起重机、推土机等，包括承包人服

English	Chinese
contractors, to replace or repair the track right-of-way and associated structures; and 2) Costs associated with the repair or replacement of roller bearings on units that were derailed or submerged in water (replacement costs means the labor costs resulting from a wheel set change out).	务，更换和修复轨道通过权和相关构造物；2）修理或更换出轨或淹没的装置上的滚动轴承的相关费用（更换成本是指因更换车轮设备所产生的劳动力成本）。
Reportable Death, Injury or Illness: Any event arising from the operation of a railroad which results in: 1) death to one or more persons; 2) injury to one or more persons, other than railroad employees, that requires medical treatment; 3) injury to one or more employees that requires medical treatment or results in restriction of work or motion for one or more days, one or more lost workdays, transfer to another job, termination of employment, or loss of consciousness; or 4) any occupational illness of a railroad employee, as diagnosed by a physician.	**可报告的死亡或伤病**：由于铁路的操作不当引起的任何事件，导致以下结果：1）死亡一人以上；2）受伤一人或多人（不包括铁路工作人员）并需要治疗；3）一名或多名铁路工作人员受伤，需要治疗或一天以上工作或活动受限，停止一天或以上的工作日，改换其他工作或中止工作，失去知觉；4)经医生诊断的铁路工作人员的任何一种职业病。
Reportable Vehicle: All sedans, station wagons, ambulances, buses, carryalls, trucks and truck tractors. Excluded are semitrailers, trailers, and other trailing equipment such as pole trailers, dollies, cable reels, trailer coaches and bodies, portable wheeled compressors, trucks with permanently mounted equipment (e.g. generators, air compressors, etc.), fire trucks, motorcycles, electric and hybrid powered electric vehicles and military design motor vehicles.	**可报告车辆**：所有私家轿车、旅行车、救护车、公共汽车、封闭式汽车、卡车、卡车拖车。半拖车、拖车和其它牵引设备如单极拖车、双系柱拖车、缆车、拖式长途汽车、轻便轮式压缩机、有永久固定设备的卡车（例如：发电机、空气压缩机等）救火车、摩托车、电动和混合动力电动车、军用设计汽车除外。

English	Chinese

Reporting Threshold: The level of railroad property damage, resulting from a train accident involving on-track equipment, over which a railroad company must report the accident to the Federal Railroad Administration. Reportable damages include the cost of labor and the cost of repairing (or replacing in kind) damaged on-track equipment, track, track structure, or roadbed.

报告临界点：因火车事故造成的关于轨道设备的铁路财产损失的最低程度。高于此点，铁路公司必须向联邦铁路管理局汇报事故。可报告损失包括劳动力成本和修复破坏的轨道设备、轨道、轨道构造物或路基的成本。

Representative Vehicle: A motor vehicle which represents the type of motor vehicle that a driver applicant operates or expects to operate.

代表车型：一种代表驾驶员申请人驾驶的或将要驾驶的汽车类型的车辆。

Research: Investigation or experimentation aimed at the discovery of new theories or laws and the discovery and interpretation of facts or revision of accepted theories or laws in the light of new facts.

研究：旨在发现新的理论或规律。发现和解释事实，或根据新的事实修正已被接受的理论和规律的调查和实验。

Research and Special Programs Administration (RSPA): The Administration was established formally on September 23, 1977. It is responsible for hazardous materials transportation and pipeline safety, transportation emergency preparedness, safety training, multimodal transportation research and development activities, and collection and dissemination of air carrier economic data. It includes the Office of Hazardous Materials Safety; the Office of Pipeline Safety; the Office of Research Technology, and Analysis; the Office of University Research and Education; the

研究和特别项目管理中心：管理中心正式成立于 1977 年 9 月 23 日。它负责危险物资运输和管线运输安全、交通应急准备、安全培训、多式联运研究和发展活动、收集和传播航空承运人的经济数据。它包括危险物资安全办公室、管道安全办公室、研究技术和分析办公室、大学研究和教育办公室、自动化交通办公室、研究政策和技术转移办公室、沃尔佩国家交通运输系统中心以及

English	Chinese
Office of Automated Tariffs; the Office of Research Policy and Technology Transfer; the Volpe National Transportation Systems Center; and the Transportation Safety Institute.	运输安全研究所。
Reset Device: A device whereby the brakes may be released after an automatic train control brake application.	**复位装置**：自动列车制动以后，能把刹车复原的装置。
Residential District: The territory adjacent to and including a highway which is not a business district and for a distance of 300 feet or more along the highway is primarily improved with residences.	**居民区**：邻近公路以及包括公路的非商业区、沿公路三百英尺以上距离主要改造用于居住的地域。
Residential Transportation Energy Consumption Survey (RTECS): This survey was designed by the Energy Information Administration of the Department of Energy to provide information on how energy is used by households for personal vehicles.	**居民交通能源消耗调查**：由能源部所属的能源信息管理中心设计。目的是提供拥有私人轿车的家庭是如何使用能源的信息。
Residential Vehicle: Motorized vehicles used by U.S. households for personal transportation. Excluded are motorcycles, mopeds, large trucks, and buses. Included are automobiles, station wagons, passenger vans, cargo vans, motor homes, pickup trucks, and jeeps or similar vehicles. In order to be included, vehicles must be 1) Owned by members of the household, or 2) Company cars not owned by household members but regularly available to	**家庭车辆**：美国家庭用于个人交通所使用的汽车。摩托车、机动脚踏两用车、大型卡车和公共汽车除外。它包括汽车、旅行车、面包车、厢式货车、机动房车、敞篷小型载货卡车和吉普车或相似车辆。符合如下条件之一的车辆属于家庭车辆：1）归家庭成员所有；或2）虽不是家庭成员所有但归家庭成员个人正常使用的平常

English	Chinese
household members for their personal use and ordinarily kept at home, or 3) Rented or leased for 1 month or more.	放在家里的公司车辆；或 3） 包租一个月以上的车辆。
Residual Fuel Oils: The topped crude of refinery operations, which includes No. 5 and No. 6 fuel oils, as defined in ASTM Specification D 396 and Federal Specification, VV-F-815C; Navy Special Fuel oil as defined in Military Specification MIL-F-859E including Amendment 2 (NATO symbol F-77); and Bunker C fuel oil. Residual fuel oil is used for the production of electric power, space heating, vessel bunkering, and various industrial purposes.	**残余燃料油**：精炼生产的拔顶原油，包括由美国材料实验协会规范 D396 和联邦规范 VV-F-815C 定义的第 5 号和第 6 号燃料油。在军用规范 MIL-F-859E 包括修订 2（北约象征殊 F-77）里定义的海军特殊燃料油，以及燃料舱 C 燃料油。残余燃料油用于电力生产、供暖、船舶燃料以及广泛的工业用途。
Residual Fuel Oils: The heavier oils that remain after the distillate fuel oils and lighter hydrocarbons are distilled away in refinery operations and that conform to American Society for Testing and Materials (ASTM) Specifications D396 and 975. Included are No. 5, a residual fuel oil of medium viscosity; Navy Special, for use in steam-powered vessels in government service and in shore power plants; and No. 6, which includes Bunker C fuel oil and is used for commercial and industrial heating, electricity generation, and to power ships. Imports of residual fuel oil include imported crude oil burned as fuel.	**残余燃料油**：就是馏出燃料油和更轻的碳氢化合物经过精炼和被提取之后留下的重油，它符合美国材料实验协会规范 D396 和 975。它包括一种中等粘度的残余燃料油 5 号油、用于为政府服务的蒸汽动力船舶和岸边动力工厂的海军特殊用油、以及包括燃料舱 C 燃油和用于商业和工业加热，电力生产和为船提供动力的六号油。残余的燃料油进口包括以进口的原油为燃料。
Residue: The hazardous material remaining in a packaging, including a tank car, after its contents have been unloaded to the maximum extent practicable and before the	**残渣**：在包装（包括油槽车）中的物质被尽可能地卸掉以后，还未重新装载或其有害物质和蒸气被彻底清除之前，该

English	Chinese
packaging is either refilled or cleaned of hazardous material and purged to remove any hazardous vapors.	包装中留下的有害物质。
Resource: Any personnel or property used in rendering assistance.	**资源**：任何用于协助的人员或财产。
Response Area: The inland zone or coastal zone, as defined in the National Contingency Plan (40 CFR 300), in which the response activity is occurring.	**响应区域**：在国家意外情况计划中定义的那些会发生应急活动的内陆或沿海区域。
Response Plan: The operator's core plan and the response zone appendices for responding, to the maximum extent practicable, to a worse case discharge of oil,or the substantial threat of such a discharge.	**对应计划**：操作者的核心计划以及对应的区域。确保最大限度切实可行，针对油的溢漏或者这种失控带来的威胁。
Response Resources: The personnel, equipment, supplies, and other resources necessary to conduct response activities.	**应急资源**：进行应急活动所必需的人员、装备、物资和其他资源。
Response Zone: A geographic area either along a length of pipeline or including multiple pipelines, containing one or more adjacent line sections, for which the operator must plan for the deployment of, and provide, spill response capabilities. The size of the zone is determined by the operator after considering available capability, resources, and geographic characteristics.	**应急区**：沿着一条管线或者包括多重管线的一个地理区域，它含有一个或多个毗邻线段，操作者必须为其部署并提供溢漏的处理应急能力。应急区的大小由操作者对其能力、资源、和地理特点进行考察后确定的。
Rest Site: A roadside area usually having facilities for people and/or vehicles.	**休息区**：在路边，为人们和车辆提供服务设施的区域。

English	Chinese
Restoring Feature: An arrangement on an electro-pneumatic switch by means of which power is applied to restore the switch movement to full normal or to full reverse position, before the driving bar creeps sufficiently to unlock the switch with control level in normal or reverse position.	**回复特征**：对电气开关进行设置，在驱动杆没有把开关放置在正常或相反位置之前，电力用来使开关回复正常或完全相反的位置。
Restraint Usage: (See also Mandatory Use Seat Belt Law, Manual Restraint System) Manually operated restraint systems include shoulder belts, lap belts, lap and shoulder belt combinations, or child safety seats. Automatic restraint systems include passive belts and air bag systems.	**约束使用**：（又见强制使用安全带法和手动约束系统）手动操作的约束系统包括肩部安全带、膝部安全带以及肩部膝部联合安全处，或者儿童安全座椅。自动的约束系统包括被动安全带和安全气囊系统。
Restricted Road: Public road with restricted public use.	**限制道路路段**：对公众使用有所限制的公共道路。
Restricted Speed: A speed that will permit stopping within one-half the range of vision, but not exceeding 20 miles per hour.	**限制速度**：能够在视距范围一半以内停下的速度，但不得超过 20 英里/小时。
Restriction of Work or Motion: The inability of a railroad employee to perform all normally assigned duties because of injury or occupational illness, and includes the assignment of a railroad employee to another job or to less than full time work at a temporary or permanent job.	**工作或活动限制**：铁路员工由于受伤或职业病无法履行正常职责。包括铁路员工转去其他工作或在以临时或长期工作身份进行不足全职的工作。
Retail Gasoline (Motor) Prices: (See also Gasoline) Motor gasoline prices calculated each month by the Bureau of Labor Statistics (BLS) in conjunction with the construction of the Consumer Price Index (CPI). Those	**汽油零售价格（汽车）**：（又见汽油）汽油价格每月由劳工统计局根据居民消费价格指数来计算。这些价格从代表美国 80％的城市人口的 85 个城市

English	Chinese

prices are collected in 85 urban areas selected to represent all urban consumers about 80 percent of the total U.S. population. The service stations are selected initially, and on a replacement basis, in such a way that they represent the purchasing habits of the CPI population. Service stations in the current sample include those providing all types of service (i.e., full -, mini -, and self-service).

搜集得到。选定的加油站定期更换，以代表所属人口的购买习惯。目前包括的加油站样本提供各种服务（例如：全、小、自助服务）。

Revenue Vehicle Mile: The distance in miles that a revenue vehicle is operated while it is available for passenger service.

汽车车辆收入里程：一个提供客运服务的车辆可行驶的里距。

Reverse Commuting: Movement in a direction opposite the main flow of traffic, such as from the central city to a suburb during the morning peak period.

反向通勤：同主交通流相反方向的出行。例如在早高峰时从中心城市向郊区的出行。

Ride Quality: A measure of the comfort level experienced by a passenger in a moving vehicle, including the vibration intensity and frequency, accelerations (longitudinal, transverse, and vertical), jerk, pitch, yaw, and roll.

乘坐质量：对乘客在车中舒服程度的度量，包括振动强度和频率、加速度（纵向、横向、垂直）、坡度、沥青、偏航、滚动。

Ridesharing: A form of transportation, other than public transit, in which more than one person shares the use of the vehicle, such as a van or car, to make a trip. Also known as "carpooling" or "vanpooling."

合乘：一种不同于公共交通的交通出行方式。几个人共用面包车或小汽车出行。也被称为合用。

Ridge Line: The line separating drainage basins.

分水岭：分开排水盆地的一条线。

Right of Way: (See also Controlled Access Rights-of-Way; Exclusive Rights-of-Way)

路权：（又见控制进入路权和排外路权）用于公路运输目的

English	Chinese
The land (usually a strip) acquired for or devoted to highway transportation purposes.	的土地（一般为长条地带）。
Road: See also Arterial, Expressway, Freeway, Highway, Local Streets and Roads, Roadway.	**道路**：又见主干道、快速路、高速公路、公路、街道和当地道路、道路。
Road: An open way for the passage of vehicles, persons, or animals on land.	**道路**：陆地上供车辆、人员、动物行走的公共空间。
Road Call: (See also Roadcalls for Mechanical Failure, Roadcalls for Other Reasons) Unscheduled maintenance requiring either the emergency repair or service of a piece of equipment in the field or the towing of the unit to the garage or shop.	**道路呼叫**：（又见机械故障呼叫和其他原因的呼叫）。紧急抢修或者对外场设备进行服务或者将设备、车辆拖走至修理厂等不定期维修。
Road Class: The category of roads based on design, weatherability, their governmental designation, and the Department of Transportation functional classification system.	**公路等级**：基于道路设计、耐候性，政府指定和交通部对道路功能的分级。
Road Functional Classification: The classification of a road in accordance with the Bureau of Land Management (BLM) 9113.16. Code as follows: C-collector, L-local, R-resource.	**道路功能分级**：依照道路土地管理局 9113.16 对道路的分类。代码如下：C-集聚道路 L-地方道路 R-资源
Road Gate: Gate blocking entrance to a road.	**路障**：阻拦门，限制人或车进入道路。
Road Hog: Motorist who takes more than his share of the highway.	**路霸**：使用道路多于其份额的驾驶员。
Road Miles: The length in miles of the single or first main track measured by the distance between terminals or stations, or both. Road	**路英里**：以英里计数的用终点之间、站点之间，或者终点和站点之间的距离衡量的单一或

English	Chinese

miles does not include industrial and yard tracks, sidings, and all other tracks not regularly used by road trains operated in such specific service, and lines operated under a trackage rights agreement.

主干道的长度。路英里不包括工业和围场轨道、房屋板壁，和所有其它的非常规的轨道，包括被这些特殊服务占用的线路以及铁道线路经营权协议下的线路。

Road Oil: Any heavy petroleum oil, including residual asphaltic oil used as a dust palliative and surface treatment on roads and highways. It is generally produced in six grades, from 0, the most liquid, to 5, the most viscous.

路油：用于道路和公路上的尘土和路面处理的任何重型油（包括石油沥青等残余）。分为六个等级，从 0、最液化、到 5、最具黏性。

Road or Trail Restrictions: Limitations placed on the use of a road or trail. Code as follows: S-seasonal closure, Y-closed yearlong to motorized vehicles, R-restriction on types of traffic allowed on road, L-limitations on vehicle dimensions, weight, or speed, N-no restrictions applied, B-no bicycles, E-no equestrians, M-no motorized vehicle, P-permit required for use.

道路和步道的限制：在道路和步道上的限制信息。代码如下：S - 季节性关闭；Y - 全年对汽车关闭；R - 对道路上允许的交通类型的限制；L - 限制车辆大小、重量或车速；N - 无限制；B - 禁止自行车；M - 禁止汽车；P - 需要使用许可证。

Roadbed: 1) In railroad construction, the foundation on which the ballast and track rest. 2) In highway construction, the graded portion of a highway within top and side slopes, prepared as a foundation for the pavement structure and shoulder.

路基：1.铁路建设中，路基用于支撑石碴和轨道；2.在公路建设方面，路基用于支撑路面结构及路肩。

Roadcalls For Mechanical Failure: (See also Road Call) A revenue service interruption caused by failure of some mechanical element of the revenue vehicle.

机械故障呼叫：由服务车辆的机械因素造成的收费服务中断。机械故障包括空气设备中断、刹车、部件、车门、冷却

English	Chinese
Mechanical failures include breakdowns of air equipment, brakes, body parts, doors, cooling system, heating system, electrical units, fuel system, engine, steering and front axle, rear axle and suspension, and torque converters.	系统、加热系统、电子元件、燃料系统、发动机、前桥、后桥、变距器。
Roadcalls For Other Reasons: (See also Road Call) A revenue service interruption caused by tire failure, farebox failure, wheel chair lift failure, air conditioning system, out of fuel-coolant-lubricant, and other causes not included as mechanical failures.	**其他原因的呼叫**：由轮胎故障、收费箱故障、轮椅升降机故障、空调系统、冷却润滑出油等非机械故障引起的收费服务中断。
Roadway: See also Road.	**道路**：见道路。
Roadway: The portion of a highway, including shoulders, for vehicular use.	**道路**：供车辆使用的路的一部分，包括路肩。
Roadway: That part of a trafficway used for motor vehicle travel.	**道路**：供汽车行驶的路的一部分。
Roadway Element: That portion of the roadway apparatus of automatic train stop, train control, or cab signal system, such as electric circuit, inductor, or trip arm to which the locomotive apparatus of such system is directly responsive.	**道路要素**：道路组成包括自动列车停车装置、列车控制、机车信号系统，如：电路、电感、及对该系统的机车直接响应的定捆杆。
Roadway Function Class: The classification describing the character of service the street or highway is intended to provide.	**道路功能分级**：对道路所能提供服务的特征做分级。
Rock It: To free vehicle from mud or snow by alternately driving forward and reverse.	**摇摆**：来回驾车前进或后退使其从泥或雪中出来。

English	Chinese
Rocker Link: That portion of an interlocking machine which transmits motion between the latch and the universal link.	**摇杆连接**：连锁机器的一部分，用来传递和链接。
Rocky Mountain Double: A combination vehicle consisting of a tractor, a 45 to 48 foot semitrailer and a shorted 28 foot semitrailer.	**落基山双挂**：由拖拉机组成的合成车辆，一个拖车，一个45到48尺半挂车和28尺半挂车。
Roll and Rest: When a long haul driver drives and stops at regular intervals to sleep.	**驾驶和休息**：长途驾驶司机驾驶车辆，在规定站休息。
Rolling Equipment: Includes locomotives, railroad cars, and one or more locomotives coupled to one or more cars.	**轧制设备**：包括机车、铁道车辆、机车以及一个或多个的汽车。
Rollover: (See also Accident, Jackknife) Rollover is defined as any vehicle rotation of 90 degrees or more about any true longitudinal or lateral axis.	**翻车**：.指任何车辆绕纵向或横向轴线旋转90度或以上。
Rolltop: Trailer with a sliding roof to permit crane loading.	**卷顶车**：顶部带有滑盖，可以用来起重的车辆。
Rotation: The reassignment of vehicles either within or between agencies to equalize mileage.	**轮换**：车辆部门或机构之内或之间分配以达到里程均等。
Route: The course or way which is, or is to be, traveled.	**路线**：行走的路线。
Route: A designated path through a road network.	**路线**：在路网中指定的通道。
Route Locking: Electric locking, effective when a train passes a signal displaying an aspect for it to proceed, which prevents the	**锁线**：电子锁定，当列车通过时一个信号显示允许它前行时时效，可以防止车辆偏移，动

English	Chinese

movement of any switch, movable-point frog, or derail in advance of the train within the route entered. It may be so arranged that as a train clears a track section of the route, the locking affecting that section is released.

点和脱轨。这样的安排可以使列车离开路段时，锁定区解锁。

Route Miles: The total number of miles included in a fixed route transit system network.

道路英里：包括在一个固定路线的公交系统网络的总英里数。

Route Segment: A part of a route. Each end of that part is identified by a continental or insular geographical location; or a point at which a definite radio fix can be established.

路段：路线的一部分。路段的两端用陆地或封闭的地理位置识别；或者在这一点设立无线电点。

Rulemaking (Regulations): The authority delegated to administrative agencies by Congress or State legislative bodies to make rules that have the force of law. Frequently, statutory laws that express broad terms of a policy are implemented more specifically by administrative rules, regulations, and practices.

制定条例：由国会或州立法机构授予行政机构制定的规则，具有法律效力。表达政策的法定法律通过更具体的行政法规、规章、习俗得到落实。

Runaway Truck Ramp: A short inclined roadway constructed of sand or other unconsolidated material that exits gradually from and generally runs adjacent to the right lane of a descending highway, expressly for the purpose of stopping runaway trucks.

卡车失控匝道：紧接于下降公路右侧的由松散砂或其他材料铺成的用于卡车失控停车的一小段倾斜道路。

Running Clearance: The distance from the surface on which an automobile is standing to the lowest point on the automobile, excluding unsprung weight.

行驶清理距离：从一辆汽车所停的地面到汽车的最低点的距离，不包括外挂重量。

English	Chinese
Running Track: A track providing end-to-end line continuity and used for working regular trains between stations or places indicated in tariffs as independent points of departure or arrival for the conveyance of passengers or goods.	**跑道**：提供端到端路线的连接和用于车站之间火车独立运送出入境旅客或货物的轨道。
Rural: Usually refers to areas with population less than 5,000.	**乡村**：通常指人口不到五千人的地区。
Rural Area: Outside the limits of any incorporated or unincorporated city, town, village, or any other designated residential or commercial area such as a subdivision, a business or shopping center, or community development.	**乡村地区**：任何市，镇，村，或任何其他指定的居住或商业领域（比如一个企业或购物中心或社区）以外的地区。
Rural Area: Include all areas of a state outside of the Federal Highway Administration (FHWA) approved adjusted census boundaries of small urban and urbanized areas.	**乡村**：包括联邦公路管理局批准的都市化地区以外的所有地区。
Rural Arterial Routes: Those public roads that are functionally classified as a part of the rural principal arterial system or the rural minor arterial system as described in volume 20, Appendix 12, Highway Planning Program Manual.	**乡村主干线**：那些按照功能划分为乡村基本干线或者次级干线的组成部分的公共道路，如公路规划纲要手册 20 卷附录 12 所描述。
Rural Highway: Rural highway is any highway, road, or street that is not an urban highway.	**乡村高速公路**：乡村高速公路是指那些非城市高速公路的高速公路、公路或街道。
Rural Major Collector Routes: Those public roads that are functionally classified as a part of the major collector	**乡村主要联结线路**：按功能划分为乡路联结系统之子类，--主要联结线路，如公路规划纲

English	Chinese
subclassification of the rural collector system as described in volume 20, Appendix 12, Highway Planning Program Manual.	要手册 20 卷附录 12 所描述。

Saddle-Mount: A device, designed and constructed as to be readily demountable, used in driveaway-towaway operations to perform the functions of a conventional fifth wheel.

托架：一种为了方便拆卸而设计和建造的装置，用作完成拖车作业时常规的额外接轮。

Saddle Tank: Fuel storage area on a tractor.

鞍形油箱：在牵引车辆上用于燃料储藏的地方。

Safety Defect: A defect in a product subject to the provisions of the 46 U.S.C. Chapter 43, which creates a substantial risk of personal injury to the public. The defective part or area may be under the provisions of 46 (U.S.C.) Chapter 43, and if so regulated, may or may not be in compliance with that standard.

安全缺陷：受到 U.S.C 第 43 章 46 条的各项条款规定产品中有缺陷的部分或者地方有可能不可以用 U.S.C 第 43 章 46 款的各项条款来规定，缺陷的部分可能符合，也可能不符合该项标准。

Sales-Weighted Miles Per Gallon: Calculation of a composite vehicle fuel economy based on the distribution of vehicle sales.

销售额加权的每加仑英里数：基于车销售额的分布来计算的车用燃料的综合经济性。

SAR Facility: A regular Reserve operated or augmented, or Auxiliary operated Coast Guard unit, such as an air station, small boat station, base (or support center), group/section, or other shore unit which has search and rescue as a primary mission.

搜索与营救设施：一种常规储备或扩大储备海岸警卫设施，或辅助性海岸警卫设施，例如机场、小船停靠站、基地(或支持中心)、海边组织/部门或其它的海岸单位等，或其它一些以搜索和营救为主要目的海岸设施。

Scheduled Service: A scheduled commercial passenger vehicle trip. The scheduled trip is generally offered at preestablished times between designated locations.

定期行程：一种定期的商业客车行程。这种定期行程通常是在预定的时间，在指定的地点之间提供。

English	Chinese
School and Other Nonrevenue Bus: (See also Intercity Bus, Motor Bus, Motorbus, Transit Bus) Bus services for which passengers are not directly charged for transportation, either on a per passenger or per vehicle basis.	**校车和其他非盈利性公共汽车**：(参见城市公车、大客车、公共汽车、捷运公车)不直接向旅客按人或车收取费用，或者以每旅客或车辆为基础收取运输费用的客车服务。
School Bus: See also Public School Transportation.	**校车**：参见公共教育运输。
School Bus: A passenger motor vehicle which is designed or used to carry more than 10 passengers in addition to the driver, and which the Secretary [of Transportation] determines is likely to be significantly used for the purpose of transporting preprimary, primary, or secondary school students to such schools from home or from such schools to home.	**校车**：能够运送十人（除驾驶员之外）以上的客运车辆，运输部长决定这种车辆主要应用于运送在家和学校之间往返的小学生、初中生。
School Bus: Includes county school buses, private school buses, and buses chartered from private companies for the express purposes of carrying students to or from school and/or school-related activities.	**校车**:包括乡村公校校车、私人学校校车和从私营企业租赁的公共汽车。主要从事运送学生往返于学校及从事与学校有关的活动的快速运输工具。
School Bus Operation: The use of a school bus to transport only school children and/or school personnel from home to school and from school to home.	**校车运营**：仅用作运送学生和学校工作人员往返于家庭和学校之间的校车的运行。
School Bus Related Crash: Any crash in which a vehicle, regardless of body design, used as a school bus is directly or indirectly involved, such as a crash involving school children alighting from a vehicle.	**校车有关事故**：无论车身设计，任何直接或间接与作为校车使用的车辆的碰撞事故。例如学生下车时发生的事故。

School Bus Service: The operation of buses exclusively to carry school passengers to and from their schools.

校车服务：专门运送学校人员往返于学校的运输服务。

School Buses: Establishments primarily engaged in operating buses to transport pupils to and from school. School bus establishments operated by educational institutions should be treated as auxiliaries.

校车公司：主要从事客车运送学生往返学校的机构。由教育机构经营的校车公司是辅助性单位。

Scrappage Rate: As applied to motor vehicles, it is usually expressed as the percentage of vehicles of a certain type in a given age class that are retired from use (lacking registration) in a given year.

报废比率：通常表示为在给定年份中，一定机动车车龄的某种车型报废退出使用(未注册)的百分比。

Search: The effort expended to locate a distressed unit by a reporting unit in terms of time and distance.

搜寻：以时间和距离为报告单位来衡量定位被困单元所作的努力。

Seating Position: The location of the occupants in the vehicle. More than one can be assigned the same seat position; however, this is allowed only when a person is sitting on someone's lap.

座位位置：交通工具中乘员乘坐的位置。只有在一个人坐在另一个人的膝盖上时，一个座位才可以安排一个以上人数。

Seating Reference Point (SGRP): The unique design H-point, as defined in Society of Automotive Engineers (SAE) J1100 (June 1984), which: 1) Establishes the rearmost normal design driving or riding position of each designated seating position, which includes consideration of all modes of adjustment, horizontal, vertical, and tilt in a vehicle; 2) Has X, Y, and Z coordinates, as defined in SAE J1100 (June 1984),

座位参考点(SGRP)：由汽车工程师学会(SAE)在 1984 年 6 月制定的 J1100 规范所定义的 H 特定设计点。它 1)建立了每一个用来驾驶或乘坐的特定座位的最靠后的正常位置。这个位置考虑了各种可能的水平、竖向、或倾斜调整模式。2)具有如同汽车工程师学会 J1100 中规定的 X,Y,Z 三维坐标值，

English	Chinese

established relative to the designed vehicle structure; 3) Simulates the position of the pivot center of the human torso and thigh; and 4) Is the reference point employed to position the two-dimensional drafting template with the 95th percentile leg described in SAE J826 (May 1987), or, if the drafting template with the 95th percentile leg cannot be positioned in the seating position, is located with the seat in its most rearward adjustment position.

从而确定了与所设计的车体的相对关系。3)模拟了人体躯干和大腿之间的连接和相对旋转轴的位置；4)用来放置由汽车工程师学会 J826 规范确定的、适用于第 95 百分位腿的二维初制模板的位置。如果该模板不能与作座位位置重合，座位参考点就是座位所能调的最靠后位置所确定的点。

Second In Command: A pilot who is designated to be second in command of an aircraft during flight time.

副驾驶：在飞行中，对飞机具有仅次于正驾驶的控制权的飞行员。

Section 15: See National Transit Database.

第 15 部分：参见国家公共交通数据库。

Section Modulus: A measure of the strength of frame side rails, determined by the cross-section area and shape of the side rails. Section modulus is not affected by the material from which the side rail is made, only by the shape and position of the rail.

截面模量：度量车架纵边梁的强度标准，大小由横截面积和纵梁的形状共同决定。截面模量不受纵梁材料的影响，仅仅受纵梁的形状和位置的影响。

Sectionalizing Switch: A switch for disconnecting a section of a power line from the source of energy.

分段开关：用于断开动力线与能量源间连接的装置。

Semi: (See also Semitrailer, Tractor-Semitrailer, Truck) Semitrailer, used loosely in reference to tractor and semitrailer unit.

半拖车:（参见拖拉机半拖车，牵引拖挂车、大货车），广义上指的是货运拖拉机车头及其拖斗。

English	Chinese

Semitrailer: See also Motor Vehicle, Tractor-Semitrailer, Truck.

动力半拖车：参见机动车、牵引拖挂车、大货车。

Semitrailer: Any motor vehicle, other than a pole trailer, which is designed to be drawn by another motor vehicle and so constructed so that some part of its weight rests upon or is carried by the self-propelled towing vehicle.

半自动拖车：除了单轴拖车外，需要由另一辆牵引车牵引、并将部分自重搭挂在该牵引车上的机动车辆。

Semitrailer: Truck trailer equipped with one or more axles and constructed so that the front end rests upon a truck tractor.

货运拖挂车：货运拖车有单轴或多轴的，前端搭挂在牵引车上的挂车。

Serious Injury: An injury that results in the amputation of any appendage, the loss of sight in an eye, the fracture of a bone, or the confinement in a hospital for a period of more than 24 consecutive hours.

严重伤害：任何导致身体肢体截除、眼睛失明、骨折或者持续住院超过 24 小时的伤害。

Serious Injury: Any injury which 1) Requires hospitalization for more than 48 hours, commencing within 7 days from the date the injury was received; 2) Results in a fracture of any bone except simple fractures of fingers, toes, or nose; 3) Involves lacerations which cause severe hemorrhages, nerve, muscle, or tendon damage; 4) Involves injury to any internal organ; 5) Involves second or third-degree burns, or any burns affecting more than 5 percent of the body surface.

严重伤害：任何满足下列条件的伤害：1) 从受伤之日起，七天内需要住院治疗四十八小时以上；2)导致除了手指、脚趾、或鼻子骨裂之外的任何骨折；3)包含引起大出血、神经、肌肉或腱损害的钝锉伤口；4)任何的内部脏器损伤；5)含造成二级、或者三级烧伤，或者任何超过 5%身体表面的烧伤。

Serious Traffic Violation: Conviction, when operating a commercial motor vehicle, of: 1) Excessive speeding, involving any single offense for any speed of 15 miles per hour or more above the posted speed limit; 2)

严重交通违规：当驾驶一辆商用机动车辆，满足下列之一时，可以定罪：1))超速, 涉及车速超过标明限速十五英里以上所造成的单一违规行为；2)

English	Chinese

Reckless driving, as defined by State or local law or regulation, including but not limited to offenses of driving a commercial motor vehicle in willful or wanton disregard for the safety of persons or property; 3) Improper or erratic traffic lane changes; 4) Following the vehicle ahead too closely; or 5) A violation, arising in connection with a fatal accident, of State or local law relating to motor vehicle traffic control other than a parking violation. (Serious traffic violations exclude vehicle weight and defect violations.)

鲁莽驾驶,如同州和地方的法律法规的规定，包括但不仅限于完全不顾人身、财产安全，故意或肆意违规冒险驾驶商用车；3)不适当的、或突然的改变行车道；4)跟随前车太近；5)违反州或地方除泊车条例外有关机动车辆交通管理的法规，并涉及到人身伤亡。(严重违反交通规则不包括违反有关车辆重量和车辆缺陷的规定)

Service Brake: The primary mechanism designed to stop a motor vehicle.

行驶制动器：设计用来制动机动车辆的机械装置。

Set It Down: To stop quickly.

搞停：急速停止。

Shag: Small, city trailer.

拖车：小型的城市拖车。

Shake the Lights: Blinking headlights as a warning signal.

闪灯：闪烁车前灯作为一种警示。

Shared Roadway: Any roadway upon which a bicycle lane is not designated and which may be legally used by bicycles regardless of whether such facility is specifically designated as a bikeway.

共用道路：没有划分出自行车道，不管该道路是否专门指定为自行车专用与否，自行车均可使用该道路。

Sheathing: A covering consisting of a smooth layer of wood placed over metal and secured to prevent any movement.

护墙板：由平整的人造木板覆于金属件上，并加以固定使之无法移动所构成的覆盖层。

Sheep Herder: Driver with questionable ability.

新手：驾驶能力不可靠的驾驶员。

English	Chinese

Shipping Paper: A shipping order, bill of lading, manifest or other shipping document serving a similar purpose and containing the information required by 49 CFR 172.202, 172.203, and 172.204.

货运文件：发货通知、装载货物帐单、船货清单和其它具有类似功能并包含有 49CFR 172.202，172.203 和 172.204 所需要信息资料的文件。

Short Ton: A unit of weight equal to 2,000 pounds.

短吨：相当于二千磅的重量单位。

Shoulder: An area adjacent to the edge of paved runways, taxiways, or aprons providing a transition between the pavement and the adjacent surface; support for aircraft running off the pavement; enhanced drainage; and blast protection.

路肩：临近飞机跑道，滑行道或者挡板围栏边，在行人道和临近路面之间过渡的区域，可以支持飞机离开边道，并可以提高排水能力和爆炸保护能力。

Shoveling: Improper loading of freight.

野蛮装卸：不正确的装货行为。

Shut In: Closed temporarily; wells and mines capable of production may be shut in for repair, cleaning, inaccessibility to a market, etc.

停产：因为修理、清洗、市场不可达性等而造成的临时关闭具有生成能力的油井和矿山的行为。

Sick Horse: A tractor in poor mechanical condition, especially with low power.

病车：机械状况很差的牵引车，特别指马力不足。

Side Extremities: The outermost point on the sides of the vehicle. Nonstructural protrusions such as tail lights, hinges, and latches are excluded from the determination of the outermost point.

侧面极限：车辆各外侧的最远点。但像尾灯、铰链和止动销等非结构的突起不包含在外极点的计算中。

Side Facing Glazing Location: Any location where a line perpendicular to the plane of the glazing material makes an angle

侧面玻璃窗位置：玻璃窗的垂直线与机车头、尾车、或客车箱的中心线成五十度以上夹

English	Chinese

of more than 50 degrees with the centerline of the locomotive, caboose or passenger car.

角。

Side Marker Lamp (Intermediate): A lamp shown to the side of a trailer to indicate the approximate middle of a trailer 30 feet or more in length.

侧置式车箱中点标示灯（中间）：在拖挂车侧面用于指示拖挂车中部的指示灯，主要用于长度超过三十英尺长的拖挂车。

Side Marker Lamps: Lamps used on each side of a trailer to indicate its overall length.

侧置式车长标示灯：拖挂车侧面用于指示车身总长的照明灯。

Sideswipe Collision: A collision of two vehicles in which the sides of both vehicles sustain minimal engagements.

侧面擦撞：两车辆的侧面沿边小限度摩擦时发生的碰撞。

Sightseeing Operations: Special service involving the transportation of passengers assembled into a travel group by the carrier and specifically designed to service some special purpose beyond mere public transportation. Such special service is to be distinguished from service which solely contemplates expeditious service between fixed points on a fixed route according to a predetermined schedule.

游览经营：是指运输由业者组成的旅客团队、且具有特殊目的的服务，不是一般意义上的公共交通。这种特种服务有别于在一条固定线路上的两点之间，根据预先确定的时间表而进行的、专门的便捷服务。

Sign: A roadway-associated feature which provides information to people passing.

交通标识：是道路上设置的向过路人或行车者提供信息的设施。

Single Axle Weight: The total weight transmitted to the road by all wheels whose centers may be included between two parallel transverse vertical planes 40 inches

单轴载荷：是指车轮的中心包含在两相距四十英寸的车辆横段面中的所有车轮传输到道路的总载荷。联邦政府允许在州

English	Chinese
apart, extending across the full width of the vehicle. The Federal single axle weight limit on the Interstate System is 20,000 pounds.	际高速公路上单轴载荷极限为两万磅。
Single Packaging: A non-bulk packaging other than a combination packaging.	**单件包装**：有别于组合包装的散装。
Single Set Over: See Knockout Single.	**单一偏距**：见单一分离。
Single Trailer Five Axle Truck: All five-axle vehicles consisting of two units, one of which is a tractor or straight truck power-unit.	**单拖挂五轴货车**：所有由两个单元构成的五轮轴货车，其中之一是作为动力单元的牵引车。
Single Trailer Four or Less Axle Truck: All vehicles with four-or-less axles consisting of two units, one of which is a tractor or straight truck power-unit.	**单拖挂四轴及四轴以下货车**：所有由两个部分组成的四轮轴及四轮轴以下的货车，其中之一是牵引车或者叫动力单元。
Single Trailer Six or More Axle Truck: All vehicles with six-or-more axles consisting of two units, one of which is a tractor or straight truck power-unit.	**单挂六轴及六轴以上货车**：所有由两个单元组成的六轮轴及六轮轴以上货车，其中之一是牵引车，也叫动力单元。
Single Unit Four or More Axle Truck: All vehicles on a single frame with four-or-more axles.	**单体式四轴及四轴以上货车**：在单一车架上，有四轮轴或四轮轴以上的所有货车。
Single Unit Four Tire Vehicle: All two-axle, four-tire vehicles, other than passenger cars. Included in this classification are pickups, panels, vans and other vehicles such as campers, motor homes, ambulances, hearses, and carryalls. other two-axle, four-tire single-unit vehicles pulling recreational or other light trailers are included in this classification.	**单体式四轮车辆**：除轿车以外，所有两轮轴四轮式车辆。包括开敞和封闭式小吨位运货汽车、厢式车及其它如野营车、房车、救护车、灵车和客运车。其它用于牵引休闲娱乐或其它轻型挂车的两轮轴四轮单体车辆也属此类。

English	Chinese

Single Unit Six Tire Two Axle Truck: All vehicles on a single frame including trucks, camping and recreational vehicles, motor homes, etc., having two axles and dual rear wheels.

单体两轴六轮货车：具有单体车架、两轮轴和双尾轮的所有车辆。包括运货车、野营、休闲车、房车等。

Single Unit Three Axle Truck: All vehicles on a single frame including trucks, camping and recreational vehicles, motor homes, etc., having three axles.

单体三轴货车：具有单体车架和三个轮轴的所有车辆。包括运货车、野营、休闲车、房车等。

Single Unit Truck: Includes two-axle, four-tire trucks and other single unit trucks.

单体载重汽车：包括两轮轴四轮胎的货车及其他单体载重汽车。

Single-Unit Truck: A medium or heavy truck in which the engine, cab, drive train, and cargo area are all one chassis.

单体载重汽车：一种中型或重型载货汽车，其引擎、驾驶室、传动系统和装载区都在同一个汽车底盘上。

Six Banger: Six-cylinder engine.

六缸发动机：六个汽缸的发动机。

Skinnie Axle: Six axle trailer.

瘦轴：六轮轴挂车。

Skins: Tires.

外胎：轮胎。

Sleeper: Truck cab with a sleeping compartment.

带卧驾驶室：一个带有卧睡处的载重汽车驾驶室。

Slip Resistant Material: Any material designed to minimize the accumulation of grease, ice, mud or other debris and afford protection from accidental slipping.

抗滑材料：设计能够将油脂、冰、泥浆或其它异物的堆积减到最低程度、并对意外滑动提供保护的材料。

Slip-Seat: Relay operation where drivers are changed periodically, but the truck continues

轮替：在车辆从起点到目的地运行的过程中，驾驶员周期性

English	Chinese

from origin to destination.

轮换接力式操作的方式。

Slow Speed: A speed not exceeding 20 miles per hour.

缓行速度：时速不超过二十英里的速度。

Small Urban Area: Places of 5,000 to 49,999 urban population (except in the case of Maine and New Hampshire) outside of urbanized areas. As a minimum the small urban area includes any place containing an urban population of 5,000 or more as designated by the U.S. Bureau of the Census.

小城市地区：城市化地区的外围，具有人口(除缅因州和新罕布什尔州之外)在五千到四万九千九百九十九之间的地区。美国人口统计署指定的小城市地区包括市区人口不少于五千的地区。

Smoker: Tractor emitting excessive smoke from exhaust.

废气制造者：尾气排放超标准的牵引车。

Smokestack: Vertical exhaust pipe on side of cab.

烟囱：在驾驶室旁边的直立的排气管。

Snowmobile: A motorized vehicle solely designed to operate on snow or ice.

雪地车：仅仅在雪地或者冰面上行使的机动车辆。

Solar Energy: The radiant energy of the sun, which can be converted into other forms of energy, such as heat or electricity.

太阳能：太阳的辐射能。能够变为热能或电能等其它形式的能量。

Solid: A material which has a vertical flow of two inches (50 mm) or less within a three-minute period, or a separation of less than one gram (18) of liquid when determined in accordance with the procedures specified in American Society for Testing and Materials (ASTM) D 4359, "Standard Test Method for Determining Whether a Material is a Liquid or Solid."

固体：在三分钟内，垂直流动不超过两英寸(50mm)的物质，或者根据 ASTM D4359(确定某物质是液体或固体的标准测试方法)的方法测定时，每十八克中分离出的液体不超过一克的物质。

English	Chinese
Solution: Any homogeneous liquid mixture of two or more chemical compounds or elements that will not undergo any segregation under conditions normal to transportation.	**溶液**：两种或两种以上的化合物或元素形成的均匀液体混合物，在正常的运输条件下，不会发生任何分离。
Spare Parts, Supplies and Other-Net: Flight equipment replacement parts of a type recurrently expended and replaced rather than repaired or reused, supplies such as unissued fuel inventories and unissued materials and supplies held in stock, prepaid items, net investments in direct financing and sales type leases and other current assets. Spare parts may be reduced by an allowance for obsolescence to provide for losses in value.	**备用件,供应品及其他纯备用零件**：周期性消耗的飞行设备替换零件，这些零件用来替换而不是修理和重新使用。供应品是指象未取用的燃料储备和未使用的原料等，通常包括存货、预付款项目、直接融资的净投资额、销售式租赁及其他流动资产。备用件会因陈旧贬值留有补贴价值金额。
Special Fuels: Consist primarily of diesel fuel with small amount of liquefied petroleum gas, as defined by the Federal Highway Administration.	**特种燃料**：根据联邦公路总局的定义，主要是指带有少量液化石油气的柴油机燃料。
Special Purpose Terminal: A specialized terminal designed to handle one type of commodity. It is a transfer facility with little or no manufacturing, although it might include lending, separating, and similar processing activities.	**专用货运终转站**：只装卸一种商品的终转站。尽管它可能有租赁、分拣、和其它类似的运作处理，但主要是一个非生产性的转运设施。
Special Purpose Vehicle: A motor vehicle commercially designed for some purpose other than the transportation of personnel, supplies, or equipment. This definition excludes any motor vehicle designed for	**专用车辆**：作为商品车设计的、但不是用于人员、物资供给和设备的运输，而是用于特定目的的机动车。此类车不包括原来设计用作普通运输，后

English	Chinese
transport and modified locally as an expedient for meeting special needs; e.g., a pickup with a snowplow blade attached.	因某种特殊、紧急的需要而在当地改装的机动车辆。例如,在小吨位运货汽车改加雪犁刀片。
Special Use Zone: An area where distinctive types of shipping activities occur.	**专用区域**：特殊类型装运活动发生的地方。
Special Vehicle: Consists of the following types of vehicles; snowmobile, farm equipment other than trucks, dune buggy or swamp buggy, construction equipment other than trucks, ambulance, large limousine, self propelled camper and motor home, fire truck, other special vehicles.	**特殊车辆**：包含以下几种车型：雪地用汽车、除运货车以外的农场设备、沙地车、沼泽轻便车、除载重汽车以外的工程建设设备、救护车、大型豪华轿车、自行式野营车和房车、消防救火车和其它特殊车辆。
Special Warehousing and Storage: Establishments primarily engaged in the warehousing and storage of special products, not elsewhere classified, such as household goods, automobiles (dead storage only), furs (for the trade), textiles, whiskey, and goods at foreign trade zones.	**特殊仓库存储**：主要是用于储存，不在别处进行分类的特殊产品的仓贮设施。这些产品包括日用品、汽车(仅限于闲置的、或不再生产的)、用于贸易的毛皮、织物、威士忌酒、外贸特区的商品等。
Specialized Carrier: A trucking company franchised to transport articles which, because size, shape, weight, or other inherent characteristics, require special equipment for loading, unloading or transporting.	**专业承运者**：对某些商品具有特许专运权的运输公司。这些商品由于其独特的尺寸、形状、重量或其它的固有特性，需要特殊设备来进行装卸和运输。
Specific Acceleration Power: Measured in watts per kilogram. Acceleration power is divided by the battery system weight. Weight must include the total battery system.	**电池输出功率比**：按瓦/千克计量，是电池输出功率除以总的电池系统重量。

English	Chinese

Specific Energy: Measured in watt hours per kilogram. The rated energy capacity of the battery divided by the total battery system weight.

比能：按瓦时/千克计量。是电池的额定能量容量除以电池系统的总重量。

Specified Minimum Yield Strength: (See also Stress Level) The minimum yield strength, expressed in pounds per square inch, prescribed by the specification under which the material is purchased from the manufacturer.

许可最低屈服强度：（参见应力水平）以磅/平方英寸为单位的最小屈服强度，可以从材料制造商的说明书里确定。

Speed Attainable In 1 Mile: The speed attainable by accelerating at maximum rate from a standing start for 1 mile, on a level surface.

一英里可达速度：在水平路面上，从静止开始以最大加速度行驶一英里所达到的速度。

Speed Attainable In 2 Miles: The speed attainable by accelerating at maximum rate from a standing start for 2 miles, on a level surface.

二英里可达速度：在水平路面上，从静止开始以最大加速度行驶二英里所达到的速度。

Speeding: Operating at a speed, possibly below the posted limit, above that which a reasonable and prudent person would operate under the circumstances.

超速驾驶：以超过一个理智、谨慎的驾驶员在这种情况下应驾驶的速度来驾驶机动车辆，尽管该速度可能在标示限速以下。

Spin Out: Lose traction on slippery roadway.

打滑：在光滑的道路上失去摩擦或咬合力

Sport-Utility Vehicle: Includes light trucks that are similar to jeeps. Other common terms for these vehicles are sport-utility, special purpose, utility or off-the-road vehicles. They may have a four or

重型家庭客车：包括类似于吉普车的轻型卡车（箱型车）。这类车的其它的通用术语包括：家庭户外多用途车，特殊用途车，多用途耐用或越野

English	Chinese

two-wheel drive. Previously called Jeep-Like Vehicle.

车。可能四轮或两轮驱动，从前被叫作类似吉普车。

Spot Market: Buying and selling of fuel for immediate or very near-term delivery by contracts to meet peak demands.

现货市场：为了满足高峰需求、实现即时送货，依合同买卖燃料的地方。

Spot the Body: Part of a trailer.

点体：挂车部分。

Spotter: (See also Yardbird) Terminal yard driver who parks vehicles brought in by regular drivers, also a supervisor who observes the activities of drivers on the road.

泊车监管员：(参见监视人) 运输中转场上车位泊停的专用驾驶员。主要帮助过场货车停靠，并监管普通司机的行为和安全。

Spring: A place where water issues from the ground naturally.

泉井：地表自然冒水的地方。

Squealer: Also known as "tattle tale" whose proper name is tachograph. A device used in a cab to automatically record miles driven, number of stops, speed and other factors during a trip.

万能计数器：也叫"闲聊器"，其合适的名称叫转速记录器。一种安装在驾驶室里,能够自动地记录一次出行中的行驶里程、停车次数、速度及其他要素的装置。

Stack: Exhaust pipe on diesel.

排气管：柴油机上的排气装置。

Stack Car: An articulated five-platform rail car that allows containers to be double stacked.

堆叠货车箱：铰接五平台铁路车皮，允许叠放双层集装箱。

Stake Body: Truck or trailer platform body with readily removable stakes, which may be joined by chains, slats, or panels.

挡板车箱：在车体四周带有可卸式车挡板的卡车或挂车平台。这些挡板可能用链条、条板或嵌板等与车体相连。

English	Chinese
Standard Body: A body type normally furnished as a standard option by the original equipment (chassis) manufacturer (e.g., pick-up, panel, and flat bed).	**标准车体**：由最初设备(底盘)制造商提供的、作为标准选择的车体类型(例如，小型货运车、车箱封闭式小货车、和敞式车体机动或挂车)。
Standard Design Vehicle: A vehicle procurable from regular production lines. Included in this category are commercial design vehicles modified for special purpose use, if the modifications have not altered the basic design characteristics of the vehicle.	**标准设计车辆**：从普通汽车生产线获得的车辆。没有改变车辆的基本设计特征、为特殊用途而商业设计改装的车辆也包括在该类中。
Standard Industrial Classification (SIC): A set of codes developed by the Office of Management and Budget which categorizes industries into groups with similar economic activities.	**工业标准分类(SIC)**：由联邦管理预算办公室根据经济活动的相似性将工业分成不同的产业而形成的一套编码。
Standard Labor Rate: A rate calculated to approximate the total per hour cost of salaries and related fringe benefits for application against labor hours in pricing job orders.	**标准劳务费率**：评价工作招聘时，用于估计劳动时间内每小时工资成本及相应的边际收益。
Standard Metropolitan Statistical Area: (See also Central City, Metropolitan Statistical Area) A county that contains at least one city of 50,000 population, or twin cities with a combined population of at least 50,000, plus any contiguous counties that are metropolitan in character and have similar economic and social relationships.	**标准都市统计区域**：(参见中心城市，大都市统计区域)至少拥有一个五万人口的城市，或者合计人口在五万以上的孪生城市的县郡，还包括具有大都市特征并且具有相似的经济社会联系的邻接县郡。
State: A state of the United States, the District of Columbia, the Commonwealth of	**州**：泛指美国的州（邦）、哥伦比亚特区、波多黎各共同

English	Chinese
Puerto Rico, the Commonwealth of the Northern Mariana Islands, the Virgin Islands, American Samoa, Guam, or any other territory or possession of the United States designated by the Secretary [of Transportation].	体、北马里亚纳群岛共同体、维京群岛、美国萨摩亚群岛、关岛及由运输部长宣布的所有美国的其它领土和属地。
State-Designated Route: A preferred route selected in accordance with U.S. DOT "Guidelines for Selecting Preferred Highway Routes for Highway Route Controlled Quantities of Radioactive Materials" or an equivalent routing analysis which adequately considers overall risk to the public.	**州定路线**：根据美国运输部"放射性物质运输数量与控制公路路线特选指引"所选择的优先路线，或者类似的充分考虑了公众总体风险的路线分析。
State of Domicile: That state where a person has his/her true, fixed, and permanent home and principal residence and to which he/she has the intention of returning whenever he/she is absent.	**定居州**：公民拥有自己真正永久固定的房屋和主要住所的所在州，并且一旦离开，无论何时他/她都有返回该州的意图。
State Routing Agency: An entity (including a common agency of more than one state such as one established by Interstate compact) which is authorized to use state legal process pursuant to 49 CFR 177.825 to impose routing requirements, enforceable by State agencies, on carriers of radioactive materials without regard to intrastate juris-dictional boundaries. This term also includes Indian tribal authorities which have police powers to regulate and enforce highway routing requirements within their lands.	**州际定线机构**：一个有权行使与 CFR 177.825 有关、或规定的州的法律程序的实体（包括几个州根据州际协定设立的共同机构)。该机构有权规定有关放射性物质运输的路线选择要求，而无须考虑各州的实际管辖边界。各州政府会强制要求业者执行这些要求。这也包括在印地安人的土地内,拥有权力管理和强制执行公路定线要求的印地安部落当局。

English	Chinese
State Transportation Agency: The State highway department, transportation department, or other State transportation agency to which Federal-aid highway funds are apportioned.	**州运输机构**：包括共享联邦公路基金资助的州公路局、交通部和其它州的运输机构。
Static Loaded Radius Arc: A portion of a circle whose center is the center of a standard tire-rim combination of an automobile and whose radius is the distance from that center to the level surface on which the automobile is standing, measured with the automobile at curb weight, the wheel parallel to the vehicle's longitudinal centerline, and the tire inflated to the manufacturer's recommended pressure.	**静荷载半径弧**：是一个以标准车胎、车箍结合体的中心为圆心的一段圆弧，它的半径是从该圆心到轮胎与地面形成的水平面的距离。测量时要求汽车在自然载重条件下，轮面与车辆纵向中心线平行，并且轮胎压力满足生产厂家的推荐压力。
Station: With respect to intercity and commuter rail transportation, the portion of a property located appurtenant to a right of way on which intercity or commuter rail transportation is operated, where such portion is used by the general public and is related to the provision of such transportation, including passenger platforms, designated waiting areas, rest rooms and, where a public entity providing rail transportation owns the property, concession areas, to the extent that such public entity exercises control over the selection, design, construction, or alteration of the property, but this term does not include flag stops (i.e., stations that are not regularly scheduled stops but at which trains will stop to board or detrain passengers only on signal or advance notice).	**车站**:对于城市间和往返旅行轨道交通来讲，车站是附属于该类轨道交通线上的设施，被公众用作站台、特定候车区和卫生间。而提供轨道运输服务的公共机构，在车站和特许地区享有财产权，并控制了车站的选址、设计、建设和改建等。但是，这种车站不包括招呼站(即非定期计划停靠、仅仅在信号或者预先通知的情况下，火车才停靠站台上下乘客的车站)。

Steel Wheel: In rail systems, the specially designed cast or forged steel, essentially cylindrical element that rolls on the rail, carries the weight, and provides guidance for rail vehicles. The wheels are semi permanently mounted in pairs on steel axles and are designed with flanges and a tapered treat to provide for operation on track of a specific gauge. The wheel also serves as a brake drum on cars with on-tread brakes.

钢轮：在铁路系统中，专门铸造或锻造的圆柱形部件，在轨道滚动时，可以运送重物并控制轨道车辆的运行方向。钢轮被成对半永久性地安装在钢轴上，并设计有轮缘和锥形处理，以便在特定规格的轨道上运行。钢轮也可用作汽车刹车系统的闸箍，造成轮胎面与路面的摩擦而刹车。

Steering Wheel Lash: The condition in which the steering wheel may be turned through some part of a revolution without associated movement of the front wheels.

方向盘间隙：指方向盘旋转与车辆前轮的转动没有同步进行的情况。

Stem Winder: Hand-crank starter.

茎发条：手摇起动器。

Stinger-Steered Combination: A truck tractor semitrailer wherein the fifth wheel is located on a drop frame located behind and below the rear-most axle of the power unit.

刺激-引导组合：指牵引拖挂车,它的第五轮所处的位置(搭挂处)在动力牵引车尾轴后下方的下垂式拖挂架上。

Stockholder's Equity: The aggregate interest of holders of the air carrier's stock in assets owned by the air carrier.

股东资产净值：在航空运输公司拥有的资产中，由航空运输公司股东持有的资产总和。

Stockyard: An enclosed area in which livestock are temporarily kept.

牲畜场：暂时用来关养家畜的封闭区域。

Stop: As applied to mechanical locking, a device secured to a locking bar to limit its movement.

锁止：一种锁定于固定锁杆以限制其移动的装置，被用作机械锁定。

Stop-Indication Point: As applied to an automatic train stop or train control system

停靠站指示点：显示要求停车信号位置的地点，用于没有采

without the use of roadway signals, a point where a signal displaying an aspect requiring a stop would be located.

用道路信号系统的自控列车停车或控制系统。

Stop Lamps: Lamps shown to the rear of a motor vehicle to indicate that the service brake system is engaged.

停车信号灯：位于机动车辆尾部、用来表明正在刹车制动的尾灯。

Stopping Distance: The maximum distance on any portion of any railroad which any train operating on such portion of railroad at its maximum authorized speed, will travel during a full service application of the brakes, between the point where such application is initiated and the point where the train comes to a stop.

制动距离：在最大许可速度下，铁路的任何路段在任一火车刹车系统全制动时间内行驶的最大距离，即从刹车作用开始点到火车停止点间的距离。

Storage Tank: A container for storing a hazardous fluid, including an underground cavern.

贮存柜：用于储藏危险液体的容器，包括地下洞穴在内。

Straight Truck: Vehicle with the cargo body and tractor mounted on the same chassis.

单体车：货物车厢和牵引机安装在同一底盘上的车辆。

Strategic Petroleum Reserve (SPR): Petroleum stocks maintained by the Federal Government for use during periods of major supply interruption.

战略石油储备 (SPR)：联邦政府保有可用的石油储存量，以备在主要石油供应中断时期使用。

Streetcar: An electrically powered rail car that is operated singly or in short trains in mixed traffic on track in city streets.

有轨电车：在城市街道的轨道上，混合运行的单节或多节电动轨道客车。

Streetcar/Trolley: Includes trolleys, streetcars, and cable cars.

市内有轨电车/有轨电车：包括电车、有轨电车、和缆车。

English	Chinese
Stress Level: (See also Specified Minimum Yield Strength) The level of tangential or hoop stress, usually expressed as a percentage of specified minimum yield strength.	**应力水平**：(参见允许最低屈服强度)剪切或环应力水平，通常用最低许可屈服强度的百分比表示。
Strip Her: Unloading the trailer.	**卸车**：拖挂车卸载。
Subsidy: Revenues received from the United States Government for the performance of guaranteed essential air service to small communities and for losses incurred in forced service. Does not include revenues from the carriage of mail at service rates or the performance of other contractual services for the Government.	**补贴**：因为对小型社区提供有保障的基本的航空服务，并且因这种强制性的服务，造成业者营运财政损失而从美国政府那里获得的收入。但不包括为政府按服务价格提供邮递服务和其他契约服务获得的收入。
Substantial Damage: As stated in 49CFR 830.2, damage or structural failure which adversely affects the structural strength, performance, or flight characteristics of the aircraft, and which would normally require major repair or replacement of the affected component. This does not include engine failure, damage limited to an engine, bent fairings or cowling, dented skin, small punctured holes in the skin of the fabric, ground damage to rotor or propeller blades, damage to landing gear, wheels, tires, flaps, engine accessories, brakes, or wingtips.	**重大损坏**：在 49 CFR 830.2 中规定：重大损坏是指严重影响飞机的结构强度、性能或飞行特性，并且通常需要大修或对损坏元件进行替换的损坏或结构破坏。不包括发动机故障和限于发动机的损坏、通风帽或者整流罩弯曲、外壳塌陷、机体表面结构小穿孔、转轴或推进器叶片磨损、着陆滑行轮系统损害和车轮子、轮胎、摆动翼片、发动机附件、刹车系统、机翼端的损坏。
Suburban Bus: A bus with front doors only, normally with high-backed seats, and without luggage compartments or restroom facilities for use in longer-distance service with relatively few stops.	**市郊公共汽车**：仅有前门的大客车。通常安装高背靠座位，并且没有在长途旅行并很少停靠时使用的行李间和卫生间设施。

English	Chinese
Subway: 1) That portion of a transportation system that is constructed beneath the ground surface, regardless of its method of construction. 2) An underground rail rapid transit system or the tunnel through which it runs. 3) In local usage, sometimes used for the entire rail rapid transit system, even if it is not all beneath the ground surface. 4) A pedestrian underpass.	**地铁**：1)不考虑其建设方法，统指交通运输系统在地表下建造的那部分。2)地下轨道快速客运系统或该系统借以运行的隧道。3)根据地方使用习惯，即使没有完全位于地表下，也可以叫地铁。4)行人地下通道。
Sucker Brakes: Vacuum brakes.	**吸盘刹车器**：真空刹车器。
Supplies and Other Net Spare Parts: Flight equipment replacement parts of a type recurrently expended and replaced rather than repaired or reused, supplies such as unissued fuel inventories and unissued materials and supplies held in stock, prepaid items, net investments in direct financing and sales type leases and other current assets. Spare parts may be reduced by an allowance for obsolescence to provide for losses in value.	**供应品及其他纯备用零件**：包括飞行器材中通常无法加以维修和再使用的经常消耗件，象燃料和库存的其它类似的无法再循环派发使用的材料和供给，所有预先付款的事项和物品，以及在直接融资、销售式租赁及其它金融流动资产中的投入等。过时的备件（物）被允许作为正常的营运消耗而消帐。
Surface Rights: Fee ownership in surface areas of land. Also used to describe a lessee's right to use as much of the surface of the land as may be reasonably necessary for the conduct of operations under the lease.	**土地权**：对地面收费报酬的所有权。也用于描述在租契下承租人根据契约所获得的对地面尽可能多的使用权，只要这种使用仍属合理履行该契约。
Surge Pressure: Pressure produced by a change in velocity of the moving stream that results from shutting down a pump station or pumping unit, closure of a valve, or any other blockage of the moving stream.	**水锤（冲击）压力**：流体速度的变化而产生的压力，这种变化是由关闭泵站或泵体单元，阀门，或者其它任何能造成流体停止流动的行为。

English	Chinese
Surplus: Any excess personal property not required for the needs and the discharge of the responsibilities of any Federal agency, as determined by the Administrator of General Services.	**盈余**：满足必需后多余并且对任何联邦机构没有责任（债权）的个人财产，通常由公共服务主管确定。
Surplus Energy: Energy generated that is beyond the immediate needs of the producing system. This energy may be supplied by spinning reserve and sold on an interruptible basis.	**剩余能源**：超出生产系统当下需要的能量。可以间断性地通过消减储备来出售这些剩余能源。
Swamper: A helper who rides with driver.	**帮手**：随车帮助驾驶员的助手。
Swindle Sheet: Interstate Commerce Commission (ICC) log.	**报销单**：州际商业委员会(ICC)记录。
Switch-and-Lock Movement: A device, the complete operation of which performs the three functions of unlocking, operating and locking a switch, movable-point frog or derail.	**转换 - 锁止动作**：一种完整的开关作业装置，可以完成解锁、运转和锁定开关、活动点辙叉或转辙器三项功能。
Switch Circuit Controller: A device for opening and closing electric circuits, operated by a rod connected to a switch, derail or movable-point frog.	**转换电路控制装置**:用于开启和关闭电路的装置，靠一根与开关、转辙器或者可移动点辙叉相连接的杆进行操作。
System: Total operations of a carrier or carrier grouping including both domestic and international operations.	**系统**：运输业者或者运输业者群体的所有作业，包括国内和国际作业。

English	Chinese
Tack: Short for tachograph or tachometer.	**转速表**: 转速图表或转速计的简称。
Tag: The official U.S. Government motor vehicle license plate, District of Columbia license plate, or license plate of any State, Territory, or possession of the United States.	**车牌**: 法定的美国政府机动车牌照，哥伦比亚特区的牌照或任何州、地区或任何美国领土内的牌照。
Tag Axle: A nonpowered vehicle axle that helps distribute the load.	**附属轴**：无动力轴,用来分担载荷的的车轴。
Tail Lamps: Lamps used to designate the rear of a motor vehicle.	**尾灯**：用于指示机动车后方的车灯。
Tailgate: The outlet of a natural gas processing plant where dry residue gas is delivered or redelivered for sale or transportation.	**出气口**：天然气加工厂的出口。从中出来的干燥残余气体被用来出售或者做为运输燃料。
Tailgating: Driving too closely behind the vehicle ahead.	**尾随**：过于紧随前方车辆行驶
Talus: Slopes of broken rock debris on a mountainside.	**斜坡**：在山腰上由岩石碎块组成的一个倾斜坡。
Tandem: (See also Dual Drive Tandem) Semitrailer or tractor with two rear axles.	**串联**：（见双驱动轴前后直排）：有两个后轴的半挂车或牵引车头。
Tandem Axle: Two axles operated from a single suspension.	**双轴**：两个轴安装在同一个悬架上。
Tandem Axle Weight: The total weight transmitted to the road by two or more consecutive axles whose centers may be included between parallel transverse vertical planes space more than 40 inches and not	**双轴载荷**：通过两个或两个以上的连接轴传递给地面的所有的重量，其中心与纵断面平行且距离大于 40 英寸但小于 96 英寸，其宽度延伸到和车辆的

English	Chinese
more than 96 inches apart, extending across the full width of the vehicle. The Federal tandem axle weight limit on the Interstate System is 34,000 pounds.	宽度一样。 美国联邦政府规定双轴的载荷重限制在三万四千英镑以内。
Tandem Wing: A configuration having two wings of similar span, mounted in tandem.	**前后双翼**：在飞机上相似跨度的两个翼展，一前一后安装。
Tank: A structure used for the storage of fluids.	**油箱**：用来存储燃油的装置。
Tank Farm: An installation used by gathering and trunk pipeline companies, crude oil producers, and terminal operators (except refineries) to store crude oil.	**贮油厂**：管道运输和储存公司、 原油生产商、车站（除炼油厂外）用来储存原油的设施。
Tank Vehicle: Any commercial motor vehicle that is designed to transport any liquid or gaseous materials within a tank that is either permanently or temporarily attached to the vehicle or the chassis. Such vehicles include, but are not limited to, cargo tanks and portable tanks, as defined in 49 CFR However, this definition does not include portable tanks having a rated capacity under 1,000 gallons.	**运油车**：用来运送液体或气体原料的任何商业机动车 （其贮油罐永久或者暂时地连接到汽车或底盘上）。这些车辆包括但不仅限于如 49 CFR 171 所定义的货物舱和 便携式油箱。但是不包括容积在一千加仑以下的便携式油箱
Tanker: See also Barge.	**油轮**：见驳轮。
Tanker: Truck, semitrailer, or trailer with a tank body for hauling liquids.	**运油车**：运送燃油的卡车、挂车、 拖车。
Tariff: A published volume of rate schedules and general terms and conditions under which a product or service will be supplied.	**关税**：公布的关税税率宗卷，产品或服务所适用的一般条款和情况。

English	Chinese
Tarp: Tarpaulin cover for open top trailers.	**防水布，油布罩**：在敞棚拖车上加盖的防水布罩。
Tattle Tale: Tachograph.	**闲话器**：转速图表
Taxi: The use of a taxicab by a driver for hire or by a passenger for fare. Also includes airport limousines. Does not include rental cars if they are privately operated and not picking up passengers in return for fare.	**出租汽车**：用于出租给司机使用或赚取乘客车费的汽车。包括机场的豪华客车。不包括私人运营的不以赚取乘客车费盈利的汽车。
Taxicabs: Establishments primarily engaged in furnishing passenger transportation by automobiles not operated on regular schedule or between fixed terminals. Taxicab fleet owners and organizations are included, regardless of whether drivers are hired or rent their cabs or are otherwise compensated.	**出租车公司**：不按照常规固定时间和路线运送乘客的汽车公司，包括所有业主和组织者，无论司机是被雇佣的还是租用汽车或用其他方式。
Taximeter: A mechanical or electrical device in a taxicab that records and indicates a charge or fare calculated according to distance traveled, waiting time, initial charge, number of passengers, and other charges authorized by ordinance or by rule. Some taximeters are part of electronic dispatching systems.	**计程器**：出租车里用来记录和提示车费的机械式或电子的装置。它是根据行车路程、等待时间、起步价、乘客人数以及有关法令和法规所收取的费用。一些计程器是电子寻呼分派系统的一部分。
Technical Factory Visit: A visit of officer in charge marine inspection (OCMI) personnel to a manufacturing facility to check for compliance with standards and regulations, examine products and answer technical questions.	**工厂技术指导检查**：海军质量检测主管部门（OCMI）人员走访生产基地，检查产品是否根据有关标准和规定进行生产并就技术上的问题进行解答。

English	Chinese
Technology Transfer: Those activities that lead to the adoption of a new technique or product by users and involves dissemination, demonstration, training, and other activities that lead to eventual innovation.	技术转让：那些使用户采用某一新科技或产品所作出的活动，包括产品发布、展示、培训以及其他导致最终创新的活动。
Temporary Living Quarters: A space in the interior of an automobile in which people may temporarily live and which includes sleeping surfaces, such as beds, and household conveniences, such as a sink, stove, refrigerator, or toilet.	临时休息室：汽车里用来供乘客进行短暂休息的空间，包括睡觉的设施如床；以及日常生活设施如水龙头、火炉、电冰箱和厕所等。
Terawatt Hour (TWH): One trillion watt hours.	兆瓦时：一兆瓦时
Terminal: Any location where freight either originates, terminates, or is handled in the transportation process; or commercial motor carriers maintain operating facilities.	节点：货运的起始站、终点站或者运输过程的中间站，或者商业运输商进行维修操作的地点。
Terminal and Joint Terminal Maintenance For Motor Freight Transport Facility: Establishments primarily engaged in the operation of terminal facilities used by highway-type property carrying vehicles. Also included are terminals which provide maintenance and service for motor vehicles.	节点和枢纽的货物运输维护站：主要服务于为公路货运车辆服务的车站或枢纽，也包括那些为车辆提供维修和服务业务的车站。
Terminal and Service For Motor Vehicle Passenger Transportation Facility: Establishments primarily engaged in the operation of motor vehicle passenger terminals and of maintenance and service facilities, not operated by companies that also furnish motor vehicle passenger transportation.	公路客运枢纽服务站：给乘客提供各种服务的客运枢纽服务站点，它们基本不由那些提供乘客运输车辆的公司操作。

English	Chinese
Terminated Carload: A carload which ends its journey and is unloaded on a particular railroad.	**终到货物**：到达终点并在指定地点转到某一铁路的货物
Terrace: A steplike feature between higher and lower ground; a relatively flat or gently inclined shelf of earth, backed and fronted by steep slopes or manmade retaining walls.	**梯田地**：有落差的像梯子似的地面，相对平坦但又微有点倾斜的地面，在陡坡地前或坡地后，也可以是人造的阶梯地面。
Territorial Highway System (THS): The full name is Federal-aid Territorial Highway System. A system of arterial and collector highways, plus inter-island connectors that are established under 23 U.S.C. 215 by each territory (Guam, Northern Marianas Islands, Samoa, and the Virgin Islands).	**美国领地公路系统（THS）**：全称是联邦资助的领地公路系统。它是一个由主干公路，联络道路和连岛公路组成的系统。这是一个在 23 U.S.C. 215 工程下由各个地方政府修建而成的（关岛、北马里亚纳群岛、萨摩亚群岛和维尔京群岛）
Test Procedures: Specifies the methods and equipment the Coast Guard uses in determining whether boats comply with applicable standards.	**检验过程**：海岸警卫队员检验船只是否符合规定的标准所使用的具体方法和设备。
Thermal Limit: The maximum amount of power a transmission line can carry without suffering heat-related deterioration of line equipment, particularly conductors.	**热度极限**：电线可以传输的不致于引起线路设备发热老化，尤其是导体损耗的最大电量。
Thermal Storage: Storing heat for use at a later time. For example, ceramic bricks can be charged up to 1,200 degrees Fahrenheit in an 8-hour period and the heat released over the next 16 hours.	**蓄热器**：备用的热能储存，例如，瓷砖在八小时内可以加热到一千二百摄氏度，所充热量可释放十六小时。

Thermosiphon System: A solar collector system for water heating in which circulation of the collection fluid through the storage loop is provided solely by the temperature and density difference between the hot and cold fluids.

温差循环系统：太阳能热水系统。在这一循环中液体的循环动力仅来自于热冷空气的对流时温度和密度的差异。

Third Party: (See also Accident) When referring to motor vehicle accidents; the Government being the first party and the government owned vehicle (GOV) operator being the second party, the third party is the other concern in an accident.

第三方：（同见事故）指车辆交通事故的第三方。其中，如果政府是第一方，政府拥有的车辆的驾驶员就是第二方，第三方指与这次事故有关的其他人或物。

Third Rail: An electric conductor, located alongside the running rail, from which power is collected by means of a sliding shoe attached to the truck of electric rail cars or locomotives.

第三轨: 电动传导轨, 安装在运行铁轨旁边。电力通过安装在电车或机车上的滑瓦传输。

Third Structure Tax: Any tax on road users other than registration fees or fuel taxes.

第三种结构税收：对道路使用者征收的除注册费和燃油税以外的税款。

Throughput: Actual or estimated volume of natural gas that may be carried on a pipeline over a period of time.

吞吐量：一段时间内通过管道所运输天然气的实际或估计总量。

Tie Line: A transmission line connecting two or more power systems.

电线：连接两个或更多动力系统的导线。

Time Release: A device used to prevent the operation of an operative unit until after the expiration of a predetermined time interval after the device has been actuated.

限时释放器：用于防止设备超时运转的设备。

English	Chinese
Time Zone: A geographic region within which the same standard time is used.	**时区**：采用相同时间标准的一个地理区域。
Timed Transfer System: A transit network consisting of one or more nodes (transit centers) and routes or lines radiating from them. The system is designed so that transit vehicles on all or most of the routes or lines are scheduled to arrive at a transit center simultaneously and depart a few minutes later; thus transfers among all the routes and lines involve virtually no waiting. TTS is typically used in suburban areas and for night service; in other words, for those cases in which headways are long.	**计时换乘系统**：包括一个或一个以上的公交节点（中心）和连接路线的公交系统。系统被设计好，所有的或大多数路段上的车辆同时抵达一个公交中心，并在之后的几分中后离开，使所有路线上的换乘乘客不需要等待。这种系统通常用于郊区或夜间交通，这种做法针对那些发车间距较大的情况。
Timing Relay: A relay which will not close its front contacts or open its back contacts, or both, until the expiration of a definite time interval after the relay has been energized.	**延时继电器**：继电器不闭合其前触点或打开后触点，或者同时闭合其前触点和打开后触点，直到在给定的时间间隔终止,继电器被激化。
Tipple: A central facility used in loading coal for transportation by rail or truck.	**煤场**：用来装载用火车或汽车运送的煤的中央设施。
Tire Capacity: The rated capacity in pounds that the tire is designed to support, as established by the current Tire and Rim Association ratings.	**轮胎承载能力**：轮胎所能承受的最大载重量（单位：磅），它是根据当今轮胎轮缘协会的等级标准而制定的。
Toll Road: Travel fee is collected at entry or exit.	**收费公路**：在公路的入口或出口进行收费的公路。
Ton Mile: (See also Average Length of Haul) One ton moved one mile.	**吨英里**：（见货物平均运送里程）度量单位,等于将一吨货物运送一英里。

English	Chinese
Ton Mile: One ton (2,000 pounds) transported one statue mile (5,280). Ton-miles are computed by multiplying the aircraft miles flown on each inter-airport hop by the number of tons carried on that hop.	**吨英里**：度量单位，将一吨货物（两千英镑）运送一英里的度量单位。吨英里是用飞机在各个机场之间的飞行距离乘以在该距离内所运输的吨数来计算。
Ton Miles Tax: A tax calculated by measuring the weight of each truck for each trip. The gross weight is assigned a tax rate which is multiplied by the miles of travel.	**吨英里税**：根据卡车每行程的载重量所收取的税款。根据车辆总重来确定相应的税率然后再乘以相应的里程数（单位为英里）
Tonne: See Metric Ton.	**吨**：见公吨。
Tooling Down the Highway: Driving vehicle along at normal speed.	**正常行驶**：车辆以正常的速度行驶。
Top Shell: The tank car tank surface, excluding the head ends and bottom shell of the tank car tank.	**顶板**：载油车油箱的表面，不包括油箱前后端。
Torque: The amount of twisting effort exerted at the crankshaft by an engine. The unit of measure is a pound-foot which represents a force of one pound acting at right angles at the end of an arm one foot long. 1) gross torque: the maximum torque developed by an engine without allowing for the power absorbed by accessory units; 2) net torque: the torque available at the flywheel after the power required by engine accessories has been provided.	**扭矩**：发动机施加给曲轴的扭曲力。度量单位是英磅-英尺，它等于垂直施加在曲轴末端一英尺的地方一英磅的力。（1）总扭矩：发动机传递给曲轴的最大扭矩，不考虑其他零件吸收的能量。（2）有效力矩：减去被发动机其他部件吸收的能量，得到的传递给飞轮的力矩。
Torso Line: The line connecting the "H" point and the shoulder reference point as	**安全带**：根据汽车工程师协会规定的连接 H 点和乘员肩膀参

English	Chinese
defined in Society of Automotive Engineers (SAE) Recommended Practice J787g, "Motor Vehicle Seat Belt Anchorage," September 1966.	照点之间的连线。操作规程建议是 J789 克，"安克雷奇机动车安全带"，1966 年 9 月。

Total Energy: All energy consumed by end-users, including electricity but excluding the energy consumed at electric utilities to generate electricity. (In estimating energy expenditures, there are no fuel-associated expenditures for hydroelectric power, geothermal energy, solar power, or wind energy, and the quantifiable expenditures for process fuel and intermediate products are excluded.)

所有能量：被最终用户消费掉的所有能量。包括电力但不包括产生电力的用电设备所消耗的能量。（在预计的能量消耗中，不包括与燃料相关的能量消耗。如水能、地热能、太阳能、风能。可计量的加工燃料和中间产品所消耗的能量也被排除在外。）

Total Energy Consumption: The sum of fossil fuel consumption by the five sectors (residential, commercial, industrial, transportation, and electric utility) plus hydroelectric power, nuclear electric power, net imports of coal coke, and electricity generated for distribution from wood, waste, geothermal, wind, photovoltaic, and solar thermal energy.

全部能量消耗：被以下五个部分（住宅、商业、工业、交通运输、电力工业设备）消耗的石油燃料、水能、原子能、煤，以及消耗的用木材、垃圾、地热、风能、光电的、太阳能所产生的所有电能。

Total Revenue Load Factor: The percent that revenue ton-miles (passenger and nonpassenger) are of available ton-miles in revenue services, representing the proportion of the overall capacity that is actually sold and utilized.

能力利用系数：它是一个做为运营吨英里（乘客运输或者非乘客（货物）运输）所占运输能力吨英里容量百分比。它表示有多少运输能力被出售和使用。

Total Ton-Miles: The aircraft miles flown on each inter-airport hop multiplied by the tons

总吨英里：飞行里程乘以在此行程中载运的交通量（乘客运

English	Chinese
of revenue traffic (passenger and nonpassenger) carrier on that hop.	输或者货物运输）。
Tour Operators: Establishments primarily engaged in arranging and assembling tours for sale through travel agents. Tour operators primarily engaged in selling their own tours directly to travelers are also included in this industry.	**旅游承办商（旅游公司）**：主要业务是通过旅行社给旅游者安排旅游行程的收费组织。直接给观光者推销他们自己的旅游行程的公司也包括在内。
Tower: A tall framework or structure, the elevation of which is functional.	**塔**：高的建筑物或者构造物，其高度是为某种实用功能而设计的
Tracked Air Cushion Vehicle: A laterally guided vehicle that is suspended above the track by an air cushion system.	**轨道气垫车**：侧面导向的,利用气垫系统悬浮在轨道上的车辆。
Tracked Levitated Vehicle: A laterally guided vehicle that is suspended above the track by magnetic levitation.	**轨道悬浮列车**：侧面导向的,利用磁力悬浮在轨道上面的列车。
Tractor: A vehicle designed for pulling loads greater than the weight actually applied to the vehicle. The trailer on which the load is carried is connected to the tractor via the fifth wheel.	**牵引车**：用于牵引比直接施加在其上的重量更多负荷的车辆。载货拖车通常用第五个轮连接到牵引车头上。
Tractor (or Truck Tractor): The noncargo carrying power unit that operates in combination with a semitrailer or trailer, except that a truck tractor and semitrailer engaged in the transportation of automobiles may transport motor vehicles on part of the power unit.	**牵引车头（或卡车车头）**：指用来带动半拖车或拖车的不载货的动力部分。除了在用车头和半拖车运输汽车时，车头的一部分被用来承载，其余情况车头均不负载。

English	Chinese
Tractor-Semitrailer: A combination vehicle consisting of a power unit (tractor) and a semitrailer.	**半拖车卡车**：由动力车头和半拖车组成的卡车。
Traffic Accident: An accident that involved a motor vehicle that occurred on a public highway or road in the United States and that resulted in property damage or personal injury. Does not include accidents that happened in a parking lot, in a driveway, on a private road, or in a foreign country.	**交通事故**：发生在美国某条公路上由于机动车辆而导致财产损坏或者人员伤亡的事故。包括发生在停车场、私人车道、私人道路、或在其它国家的交通事故
Traffic Alert and Collision Avoidance, Type I System (TCAS): Utilizes interrogations of, and replies from, airborne radar beacon transponders and provides traffic advisories to the pilot.	**空中预警和防撞系统 I（ＴＣＡＳ）**：通过与空中雷达收发机的问讯和答复给飞行员提供飞行交通信息咨询服务的系统。
Traffic Alert and Collision Avoidance, Type II System (TCAS): Utilizes interrogations of, and replies from airborne radar beacon transponders and provides traffic advisories and resolution advisories in the vertical plane.	**空中预警和防撞系统 II（ＴＣＡＳ）**通过与空中雷达收发机的问讯和答复给飞行员在垂直方向提供飞行交通信息和策略服务的系统。
Traffic Alert and Collision Avoidance, Type III System (TCAS): Utilizes interrogation of, and replies from, airborne radar beacon transponders and provides traffic advisories and resolution advisories in the vertical and horizontal planes to the pilot.	**空中预警和防撞系统 III（ＴＣＡＳ）**通过与空中雷达收发机的问讯和答复给飞行员在垂直和水平方向提供飞行交通信息和策略服务的系统。
Traffic Assignment Zone: In planning , a division of a study area that is represented by a centroid and used for traffic assignment purposes.	**交通布局单元**：在交通规划中，将研究区域划分成单元且用一中心点来表示该区域以便进行交通布局

English	Chinese

Traffic Circle: A junction of roads that form a circle around which traffic normally moves in one direction.

交通环岛：修建成环形,车辆绕它朝一个方向行驶的交叉路口。

Traffic Control Device: A sign, signal, marking, or other device placed on or adjacent to a street or highway, by authority of a public body or official that has jurisdiction, to regulate, warn, or guide traffic.

交通控制设备：由相关政府相关管理部门或公共团体为管理、警示和指挥交通而安装在公路或街道上方或者旁边的标志物、信号灯、标线或其它设施。

Traffic Control System: A block signal system under which train movements are authorized by block signals whose indications supersede the superiority of trains for both opposing and following movements on the same track.

交通控制系统：通过发出闭塞信号来确定在同一铁轨上行驶的对向或随行列车的行车优先权的交通管制系统。

Traffic Count: A record of the number of vehicles, people aboard vehicles, or both, that pass a given checkpoint during a given time period.

交通数据：在特定的时间内，关于通过某一地点的车辆数、承载的人数或者二者之和的记录。

Traffic Inspection Facility: An area having facilities to examine pedestrian and vehicular traffic and/or cargo.

交通监管设施：一个地区所装配的用于监测行人、车辆或者货物的设施

Traffic Locking: Electric locking which prevents the manipulation of levers or other devices for changing the direction of traffic on a section of track while that section is occupied or while a signal displays an aspect for a movement to proceed into that section.

交通锁闭 ：是电子锁闭装置，用来防止对控制杆或其他设备的误操纵。这些设备是用来指示车辆在某一线路上的行驶方向，以及通过信号表示的行使许可方向的变化。

Traffic Pattern: The traffic flow that is prescribed for aircraft landing at, taxiing on,

航空线路：用来描述在某机场降落、滑行、起飞交通流量分

English	Chinese
or taking off from, an airport.	布。
Traffic Violation: See Serious Traffic Violation.	**交通违章**：见严重交通违规。
Trafficway: That part of a trafficway designed, improved, and ordinarily used for motor vehicle travel.	**道路、公路**：通常为机动车行驶而设计,改进和日常使用的道路。
Trafficway Class: A classification of highways based on a route sign.	**公路等级**：根据公路标号而对公路进行分类的一种方法。
Trail: A cleared path, beaten track, or improved surface, as through woods or wilderness, not usually trafficked by vehicles because of width, seasonal conditions, or access restrictions.	**小径**：穿过森林或者荒野的经过平整或者修理的小道。由于宽度、季节条件或者是管制限制,这些道路基本不是为机动车辆行驶为目的的。
Trailer: A motor vehicle with or without motive power, designed for carrying persons or property and for being drawn by another motor vehicle.	**拖车**：设计用来运送人或者货物的,由其他机动车牵引的,有或无动力的车辆
Trailer Converter Dolly: A trailer chassis equipped with one or more axles, a lower half of a fifth wheel and a drawbar.	**拖车连接轴**：底盘上配有一个或多个轴、下半区具有一个第五轮和一个牵引挂钩的拖车底盘 的拖车。
Trailer On a Flat Car/Container On a Flat Car (TOFC/COFC): (See also Intermodal) Often referred to as intermodal service.	**驼背运输/平板车集装箱运输（TOFC/COFC）**：（同见综合运输）通常指综合运输服务。
Trailership: A vessel, other than a carfloat , specifically equipped to carry motor transport vehicles and fitted with installed securing devices to tie down each vehicle. The term trailership includes Roll-on/Roll-off (RO/ RO)	**汽车运载船**：专门运送机动车辆并装有用来固定车辆的安全设备的轮船。这包括那些机动车可以直接驶入驶出的船只。

English	Chinese
vessels.	
Trailing Movement: The movement of a train over the points of a switch which face in the direction in which the train is moving.	**尾随行驶**：火车的运行方向与道岔的朝向相同。
Trailing Point Switch: A switch, the points of which face away from traffic approaching in the direction for which the track is signaled.	**尾部道岔**：转化器，其方向背离信号指示的车辆运行方向。
Transfer Capability: The overall capacity of interregional or international power lines, together with the associated electrical system facilities, to transfer power and energy from one electrical system to another.	**变电能力**：区域间或者国际间所有输电线以及相关电气设备，把电能从一个电力系统转变到另一个的功率总和。
Transfer Center: A fixed location where passengers interchange from one route or vehicle to another.	**交通换乘枢纽**：乘客换乘路线或者车次的固定地点。
Transfer Charge: An extra fee charged for a transfer to use when boarding another transit vehicle to continue a trip.	**换乘费**：当换乘到另一个公共交通工具继续旅行时乘客所需支付的额外费用。
Transfer Piping: A system of permanent and temporary piping used for transferring hazardous fluids between any of the following: liquefaction process facilities, storage tanks, vaporizers, compressors, cargo transfer systems, and facilities other than pipeline facilities.	**管道换装**：在下述的设备之间运送危险液体的永久的或临时的管道换装设备，包括：液化设备、储油罐、蒸馏器、压缩机、货物换装系统、以及一些非管道设施
Transfer System: Includes transfer piping and cargo transfer system.	**换装系统**：包括管道换装和货物中转系统。
Transformer: An electrical device for	**变压器**：改变交流电电压的设

English	Chinese
changing the voltage of alternating current.	备。

Transit: A partial or complete upbound or downbound passage of a vessel through one or more locks of a Seaway canal.

通过：部分或者整个船只在上行或者下行方向通过一个或者多个运河船闸

Transit Mode: (See also Rail Mode, Rapid Rail, Rapid Transit Rail, Transit Railroad, Transit Railway) Generally defined as urban and rural public transportation services including commuter trains, ferry service, heavy rail (rapid rail) and light rail (streetcar) transit systems, and local transit buses and taxis.

公交方式：（同见铁路、快速铁路车、快速铁路公交、公交铁路）一般指城市和乡村的公共交通服务包括市郊火车、轮渡、重轨铁路（快速铁路）、轻轨（有轨电车），以及地方的公共汽车和出租车

Transit System: An organization (public or private) providing local or regional multi-occupancy-vehicle passenger service. Organizations that provide service under contract to another agency are generally not counted as separate systems.

公交系统：提供本地或者地区间的多乘客运输方式服务的（政府或私人的）组织。根据合约给某些部门提供服务的通常不计为独立的系统。

Transitway: A dedicated right-of-way, most commonly in a mall, that is used by transit units (vehicles or trains), usually mixed with pedestrian traffic.

公交专用路: 最常用于商业区，通常与行人交通设施配合使用的为公交车辆（公交车或火车）所指定的专用空间。

Transmission Gas Company: A company which obtains at least 90 percent of its gas operating revenues from sales for resale and/or transportation of gas for others and/or main line sales to industrial customers and classifies at least 90 percent of mains (other than service pipe) as field and gathering, storage and/or transmission.

汽油传输公司: 从其所经营的汽油运输业务中获得至少90%的运行经费的公司。这些项目包括：汽油贩卖和再贩卖，为其他客户运输汽油，向为工业客户提供主要输送干线。公司90％以上的业务应该在聚集、储藏及[或] 传输汽油业务。

English	Chinese

Transmission Network: A system of transmission or distribution lines so cross-connected and operated as to permit multiple power supply to any principal point.

传输网络: 一个交错连接、多元控制的电力传输或分配的系统，能将多种电力供给给任何一个主要接点。

Transmission Pipeline: Pipelines (mains) installed for the purpose of transmitting gas from a source or sources of supply to one or more distribution centers, or to one or more large-volume customers, or a pipeline installed to interconnect sources of supply. In typical cases, transmission lines differ from gas mains in that they operate at higher pressures, are longer, and the distance between connections is greater.

传输管道: 为了传输从一个源头或多个补给的来源到一或较多的分配中心或是为了供给大客户，或是为互相连接供给源而安装的管道。这种传输管道和汽油管线不同，它们是在高压下传输的距离较大并且接口之间的距离大。

Transmission Type: The transmission is the part of a vehicle that transmits motive force from the engine to the wheels, usually by means of gears for different speeds using either a hydraulic "torque-converter" (automatic) or clutch assembly (manual). On front-wheel drive cars, the transmission is often called a "transaxle". Fuel efficiency is usually higher with manual rather than automatic transmissions, although modern, computer-controlled automatic transmissions can be efficient.

动力传送类型: 力矩传输是将动力从引擎传送到车轮的部份，通常由不同的速度齿轮组成（对于自动车是通过液压力矩转换器实现，对于非自动车是通过离合器实现的）。在前轮驱动汽车上，传输时常被称为"驱动桥"。手动车的能量转化效率通常比自动档的车高,尽管现代计算机控制的自动传输系统在某些情况下可以很有效率。

Transmission Types: A3-Automatic three speed, A4-Automatic four speed, A5-Automatic five speed, L4-Automatic lockup four speed, M5-Manual five speed.

动力转换类型：A3- 自动三速档,A4- 自动四速档,A5- 自动五速档, L4- 自动琐死四速档,M5- 手动五速档。

English	Chinese
Unit Load Device: Any type of freight container, aircraft container, aircraft pallet with a net, or aircraft pallet with a net over an igloo.	**成组装运装置**：任何形式的货运集装箱，航空器用集装箱，有网的航空器托盘，或有网的圆顶型航空器托盘。
Unit Tow: An integrated tow consisting of bow, center, and stem sections. Found generally in the liquid	**牵引装置**：一种包含弓，中心和主干部分的牵引装置，主要应用于流体中。
United States (U.S.) Territories: Include Samoa, Guam, the Northern Marianas, and the Virgin Islands.	**美国国土**：包括萨摩亚群岛，关岛，北马里亚纳和维京群岛。
Unlatch: Release lock on fifth wheel to drop trailer.	**解锁:** 解开第五轮锁而松开拖车。
Unloaded Vehicle Weight: The weight of a vehicle with maximum capacity of all fluids necessary for operation of the vehicle, but without cargo, occupants, or accessories that are ordinarily removed from the vehicle when they are not in use.	**卸载的车辆重量**：备有满足车辆正常运行的燃油及其他液体并达到最大负荷的车辆重量，但在车辆不使用时，通常会减掉货物，人员或其他附件的重量。
Unpaved Road Surface: Gravel/soil and unimproved roads and streets (Surface/Pavement Type Codes 20, 30 and 40).	**未铺路面**：砾石/土质和未改良的道路和街道（表面/路面类型代码 20，30 和 40）
Up-and-Down Rod: [with respect to rail operations] A rod used for connecting the semaphore arm to the operating mechanism of a signal.	**上下标杆**：[有关铁路运营] 一种用来连接信号灯臂和信号控制装置的标杆。
Upper Coupler Assembly: A structure consisting of an upper coupler plate, king-pin and supporting framework which interfaces with and couples to a fifth wheel.	**上耦合器组件**：一种包含上耦和器板，大钉和接合并连接到第五轮的支撑框架。

English	Chinese
Upper Coupler Plate: A plate structure through which the king-pin neck and collar extend. The bottom surface of the plate contacts the fifth wheel when coupled.	**上耦和器板**：通过大钉的颈部和箍圈延伸的一种面板结构。当连接后，面板的底面就会接触到第五个轮。
Upper-Half of Saddle-Mount: (See also King-Pin Saddle-Mount, Lower-Half of Saddle-Mount, Saddle-Mount) that part of the device which is securely attached to the towed vehicle and maintains a fixed position relative thereto, but does not include the "king-pin."	**鞍型装置上部**：（见大钉鞍型装置，鞍型装置下部，鞍型装置）这种装置会安全的依附于被拖车辆并保持相对固定的位置，但不包括"大钉"。
Urban Arterial Routes: Those public roads that are functionally classified as a part of the urban principal arterial system or the urban minor arterial system as described in volume 20, appendix 12, Highway Planning Program Manual.	**城市干道**：这种公共道路在功能上划为城市主干道系统或城市次级干道的一部分，其相关表述见公路规划项目手册 20 卷附录 12。
Urban Collector Routes: Those public roads that are functionally classified as a part of the urban collector system as described in volume 20, appendix 12, Highway Planning Program Manual.	**城市支路**：这种公共道路在功能上划为城市支路系统的一部分，其相关表述见公路规划项目手册 20 卷附录 12。
Urban Highway: Any highway, road, or street within the boundaries of an urban area. An urban area is an area including and adjacent to a municipality or urban place with 5,000 or more population. The boundaries of urban areas are fixed by the states, subject to the approval of the Federal Highway Administration, for purposes of the Federal-Aid highway program.	**城市公路**：城市地区分界线以内的各种公路，道路和街道。城市地区是一个包含和毗邻拥有 5000 及以上人口的城区或市区的地区。城市地区的边界是根据州政府规定而固定的，受联邦公路管理局以联邦援助公路项目为目的管制。

English	Chinese

Urbanized Area: 1) Areas with a population of 50,000 or more, at a minimum, encompass an entire urbanized area in a state, as designated by the U.S. Bureau of the Census. The Federal Highway Administration (FHWA) approved, adjusted urbanized area boundaries include the census defined urbanized areas plus transportation centers, shopping centers, major places of employment, satellite communities, and other major trip generators near the edge of the urbanized area, including those expected to be in place shortly. 2) An approximate classification of sample households as belonging to an urbanized area or not. Those classified as belong to an urbanized area were either in a central city of a Metropolitan Statistical Area (MSA), or in a MSA but outside the central city, and within a zip code area with a population density of at least 500 people per square mile in 1990.

城市化地区：１）具有五万及以上人口的地区，至少包含一个州内的由美国统计局指定的完全城市化地区。联邦公路管理局核准，调整城市化地区的分界，包括人口普查定义的城市化地区加上运输中心，购物中心，雇员集中地区，卫星社区，和其他在城市化地区边缘的交通密集和短期会增长成为交通密集的地区。２）一种样板家庭的大致分类检验是否属于城市化地区。这种城市化地区分类要么在大都会统计区的中心城市，要么在大都会统计区的中心城市之外，并在 1990 年至少具有每平方英里五百人的人口密度的邮编区以内。

User Charge: A fee charged to users for goods and services provided by the federal, state and local governments. User charges, either directly or indirectly, are paid on a periodic or occasional basis.

用户费用：由联邦，州和地方政府提供的商品及服务而向用户征收的费用。直接或间接的用户费用根可定期或不定期支付。

Utility Employee: A railroad employee assigned to and functioning as a temporary member of a train or yard crew whose primary function is to assist the train or yard crew in the assembly, disassembly or classification of rail cars, or operation of trains (subject to the conditions set forth in 49CFR218.22).

公共设施雇员：被指定作为临时的列车或调车场人员的铁路职员，主要职责为协助列车或调车场员工组装，拆卸或分类车辆，或操纵列车。（依照 49CFR218.22 条款的解释）

English	Chinese
Van: Privately owned and/or operated vans and minivans designed to carry from 5 to 13 passengers or to haul cargo.	**厢式货车**：私有的和/或者个人驾驶的货车以及设计装载五到十三人或者拉载货物的小型货车。
Vanpool: A voluntary commuter ridesharing arrangement, using vans with a seating capacity greater than 7 persons (including the driver) or buses, which provides transportation to a group of individuals traveling directly from their homes to their regular places of work within the same geographical area, and in which the commuter/driver does not receive compensation beyond reimbursement for his or her costs of providing the service.	**中型客运共乘**：一种自願式的通勤共乘安排，载客人数大于七个（包括司机）的厢式货车或者小型公共汽车，在相同地理区域内载客直接从家中出发到日常工作地点，并且通勤者和司机僅能获得补偿不超過这项旅程的花费。
Vaporization: An addition of thermal energy changing a liquid or semisolid to a vapor or gaseous state.	**蒸发**：添加热能使液体或半固体变为蒸汽或者气态。
Vaporizer: A heat transfer facility designed to introduce thermal energy in a controlled manner for changing a liquid or semisolid to a vapor or gaseous state.	**蒸馏器**：一种热转换装置,用于在控制状态下将热能导入，使液体或半固体变为蒸汽或者气态。
Vehicle: As the term is applied to private entities, does not include a rail passenger car, railroad locomotive, railroad freight car, or railroad caboose, or other rail rolling stock described in section 242 of title III of the Act.	**车辆**：这个术语用于私有个体，不包括轨道客车，铁路机车，铁路运货车，铁道守车或者其他在法案第三章 242 节所描述的轨道车辆。
Vehicle Configuration: The combination of vehicular units comprising a commercial motor vehicle.	**车辆组成**：车辆单元的组合包括一个商用机动车。

English	Chinese

Vehicle Fuel Tank Capacity: The tank's unusable capacity (i.e.,the volume of fuel left at the bottom of the tank when the vehicle's fuel pump can no longer draw fuel from the tank) plus its usable capacity (i.e., the volume of fuel that can be pumped into the tank through the filler pipe with the vehicle on a level surface and with the unusable capacity already in the tank). The term does not include the vapor volume of the tank (i.e., the space above the fuel tank filler neck) nor the volume of the fuel tank filler neck.

车辆油箱容量：油箱的不可用容量（即，当车辆油泵再也不能从油箱抽油时油箱底部剩油的体积）加上该车的可用容量（即，在不可用容量已经在油箱中时，当车在水平面时可以通过加油管注入油箱的油的体积）。这个术语不包括油箱中蒸汽的体积（即，油箱注油口以上的空间）也不包括油箱注油口的体积。

Vehicle Maneuver: (See also Crash, Vehicle Role) Last action (maneuver) this vehicle's driver engaged in either 1) just prior to the impact or 2) just before the driver realized the impending danger.

车辆操纵：（参见撞车，车辆角色）这辆车的司机最后在1）发生影响前，或者2）意识到临近的危险前所作的动作（操纵）。

Vehicle Mile of Travel (VMT): A unit to measure vehicle travel made by a private vehicle, such as an automobile, van, pickup truck, or motorcycle. Each mile traveled is counted as one vehicle mile regardless of the number of persons in the vehicle.

车辆行驶里程数：用于衡量私车行驶里程的单位，比如汽车，厢式货车，轻型货车，或者摩托车。无论车中承载的人数，每行驶一英里计作一个车辆英里。

Vehicle Miles: Vehicle miles are the miles of travel by all types of motor vehicles as determined by the States on the basis of actual traffic counts and established estimating procedures.

车辆行驶英里：车辆行驶英里是指所有类型的机动车所行驶的英里数，由该州实际交通量为基础和建立的估算程序所决定的。

Vehicle Role: (See also Crash, Vehicle Maneuver) Role of vehicle in single or multi-vehicle crashes (i.e., non-collision, striking, and struck).

车辆角色：（参见撞车和车辆操纵）车辆在单辆和多辆撞车中所扮演的角色（即，非碰撞，撞击，碰击）。

English	Chinese
Vehicle Trip: A trip by a single vehicle regardless of the number of persons in the vehicle.	**车程**：不管车辆装载的人数，单辆车的一次行程。
Vehicle Type: A series of motor vehicle body types that have been grouped together because of their design similarities.	**车辆类型**：一系列机动车车身类型，由于它们的设计相似性而被分成一组。
Very High Frequency (VHF) Communications: Provides radio voice communications between aircraft.	**高频通信**：为飞行器之间提供无线电语音通信。

English	Chinese
Walk: Includes jogging, walking, etc., provided the origin and destination are not the same	**步行**：包括慢跑，散步等等，所提供的起点与终点不一样。
Waybill: The document covering a shipment and showing the forwarding and receiving stations, the name of consignor and consignee, the car initials and number, the routing, the description and weight of the commodity, instructions for special services, the rate, total charges, advances and waybill reference for previous services, and the amount prepaid.	**货运单**：该文件涵盖船运并列出发货和接收站，发货人和收货人的名字，该车的首字母和号码，路线，描述和货物重量，特别服务的说明，比率，总费用，高级和提供服务的货运单证明信以及预付的数额。
Weekday: From 6 a.m. Monday to 5:59 p.m. Friday.	**工作日**：从星期一上午 6 点到星期五下午 5：59。
Weekend: From 6 p.m. Friday to 5:59 a.m. Monday.	**周末**：从星期五晚上 6 点到星期一早上 5：59。
Weight-Distance Tax: A tax basing the fee per mile on the registered gross weight of the vehicle. Total tax liability is calculated by multiplying this rate times miles traveled.	**重量距离税**：基于对注册的机动车总重行使一英里所需费用而征的税。计税方式为一定比率乘以行驶的里程数。
Wheelchair: A mobility aid belonging to any class of three or four wheeled devices, usable indoors, designed for and used by individuals with mobility impairments, whether operated manually or powered. A "common wheelchair" is such a device which does not exceed 30 inches in width and 48 inches in length measured two inches above the ground, and does not weigh more than 600 pounds when occupied.	**轮椅**：一种属于三轮或四轮装置的类别，可用于室内的，设计并用于行动障碍的个人的，无论手动或驱动的行动辅助工具。"通常的轮椅"就是这样一种装置，宽度不超过 30 英寸，长度不超过 48 英寸，距离地面 2 英寸，并且载荷总重不超过 600 磅。

English	Chinese
Wide Spread: Trailer axles which are more than 8 feet apart.	**宽度延伸**：拖车轴距大于 8 英尺。
WIM: Weigh-In-Motion	**不停车秤重**: 卡车，货車或拖車不須停止在高速公路旁的載重檢測站，此載重感應器可檢測仍然在公路行駛中車輛的載重。
Winch Rig: Straight truck or tractor with a hoist.	**绞车**：具有起重机械的卡车或拖车。
Windshield: The combination of individual units of glazing material of the locomotive, passenger car, or caboose that are positioned in an end facing glazing location.	**挡风玻璃**：机车，轿车或守车的玻璃材料单元的组合，位于车辆前后位置。
Winglet (Tip Fin): An out-of-plane surface extending from a lifting surface. The surface may or may not have control surfaces.	**小翼（翼尖整流罩）**：从一个上升表面延伸的机外表面。这种表面可能或可能没有控制面。
Woodchuck: Driver with low job seniority.	**土拨鼠**：低资历的驾驶员。
Wooden Barrel: A packaging made of natural wood, of round cross-section, having convex walls, consisting of staves and heads and fitted with hoops.	**木桶**：一种用天然木头制成，有圆形的横截面，有凸壁的盛具，由侧板，盖子和箍圈组成。
Work Environment: The work environment is comprised of the physical location, equipment, materials processed or used, and the activities of an employee while engaged in the performance of his work, whether on or off the railroads property. There are no stated exclusions of place or circumstance.	**工作环境**：包括工作地点，设备，处理或用过的材料，该员工从事他相关工作的活动，是否和铁路财产相关。这里没有说明的地方或情况排除。

English	Chinese
Work Equipment: Equipment which can be coupled in a train for movement over the carrier's tracks, and which is used in the carrier's work service. Includes such equipment as ballast cars, business cars, company cars, derrick cars, ditching cars, outfit cars, pile drivers, snow dozers, tool cars, wrecking cars, and others.	**工作设备：**一种可以连接在轨道运行列车上的，并可在工作服务中使用的设备。包括压舱车，商务车，公司车，起重车，开沟车，装备车，打桩机，雪推土机，工具车，抢险车等等。
Work Train: Work trains are non-revenue trains used for the administration and upkeep service of the railroad. Examples are: official trains; inspection trains; special trains running with company fire apparatus to save the railroad's property from destruction; trains that transport the railroad's employees to and from work when no transportation charge is made; construction and upkeep trains run in connection with maintenance and improvement work; and material and supply trains run in connection with operations.	**工作车：**工作列车是一种应用于管理和铁路维修的非营利性列车。例子有：官方列车;检查列车，带有公司消防设备为挽救铁路财产毁坏的特别列车；当未产生运输费用，接送铁路员工上下班的列车；在保养和维修路段中使用的工程保养车；与运行有关的物力和供应列车。
Work-Related: Any event, exposure, activity, etc., occurring within the work environment resulting in death, injury, illness to an employee is generally considered to be work-related, regardless who was responsible or at fault.	**工作有关的：**任何事件，爆炸，活动等。在工作环境中发生而导致员工死亡，受伤，疾病而不管是谁的过错，都通常被认为是工作相关的。
Worker: Any railroad employee assigned to inspect, test, repair, or service railroad rolling equipment, or their components, including brake systems. Members of train and yard crews are excluded except when assigned	**工人：**任何被指派检查，检测，维修或服务于铁路机车及零件包括刹车系统的铁路员工。除了当机车不是列车的一部分或车厂的运营被称为运营

English	Chinese
such work on railroad rolling equipment that is not part of the train or yard movement they have been called to operate (or been assigned to as "utility employees"). Utility employees assigned to and functioning as temporary members of a specific train or yard crew (subject to the conditions set forth in 49 CFR 218.22), are excluded only when so assigned and functioning.	时（或被指派为"公共设施雇员"）而指派这种工作，列车及车场员工是被排除在外的。公共设施雇员其职责是被指派为有一特定列车或车场员工的临时成员（依照49CFR218.22条款的解释）并且只有在这种指派和职责下才会被排除。
Worst Case Discharge: The largest foreseeable discharge of oil, including a discharge from fire or explosion, in adverse weather conditions. This volume will be determined by each pipeline operator for each response zone and is calculated according to 49 CFR 194.105.	**最坏情况排放**：可预见的最大的油料排放，包括火灾或爆炸及不利的天气环境的排放。排放量的多少取决于每一反应区的管线运营商，并可根据49 CFR 194.105 计算出来。
Wrecker: Truck designed for hoisting and towing disabled vehicles.	**拖车**：一种设计用来吊起和拖行故障车的卡车。

English	Chinese
Yard: A system of auxiliary tracks used exclusively for the classification of passenger or freight cars according to commodity or destination; assembling of cars for train movement; storage of cars; or repair of equipment.	**车场**：一种专门应用于客车或货车根据商品或目的地分类的辅助追踪系统；一个可以组装列车车厢，停靠列车或设备维修的场所。
Yard: A system of tracks within defined limits, whether or not part of a terminal, designed for switching services, over which movements not authorized by time table or by train order may be made, subject to prescribed signals, rules and requirements.	**车场**：在规定限度内，一系统的轨道线路，不管是否属于站场，设计用于交换服务，车辆运动不受时刻表安排,而是服从于规定的信号，规则和要求。
Yard Caboose: A caboose that is used exclusively in a single yard area.	**车场尾车**：专门用于单独车场地区的尾车。
Yard Locomotive: A locomotive that is operated only to perform switching functions within a single yard area.	**车场机车**：在单独车场地区只用于操作交换功能的机车。
Yard Mule: Small tractor used to move semitrailers around the terminal yard.	**车场骡**：用来移动车场附近半拖车的小卡车。
Yard Switching Train Mile: Computed at the rate of 6 mph for the time actually engaged in yard switching service if actual mileage is not known.	**车场转换列车里程**：在实际里程未知的情况下，以时速六英里为标准计算从事于车场转换服务的实际时间。
Yard Switching Trains: Those trains operated primarily within yards for the purpose of switching other equipment. Examples include the making up or breaking up of trains, service industrial tracks within yard limits, storing or classifying cars, and other similar operations. Switching	**车场转换列车**：这种列车主要用于场内交换其它设备。如组装或分解列车，在车场范围内服务于货车，停靠或分类列车，及其他类似业务。道路人员附带的道路执行的操作转换

English	Chinese
performed by a road crew that is incidental to the road operation is not included.	不包括在内。
Yard Track: A system of tracks within defined limits used for the making up or breaking up of trains, for the storing of cars, and for other related purposes, over which movements not authorized by timetable, or by train order may be made subject to prescribed signals, rules or other special instructions. Sidings used exclusively as passing track and main line track within yard limits are not included in the term yard track.	**车场轨道：**限定范围内的铁轨系统,用于装或分解列车，停靠列车及其他相关目的，车辆运动不受时刻表安排,而是服从车辆顺序，信号，规则和特定要求。用于超车的支线专和场内的干线铁路则不包含在车场轨道内。
Yardbird: (See also Spotter) A driver who connects and disconnects tractor and semitrailer combinations and moves vehicles around the terminal yard.	**调度员（见观察员）：**一种连接或断开牵引车或半牵引车组合并在厂内移动车辆的驾驶员。

English	Chinese
Zephyr Haul: A shipment of light weight cargo.	**和风重载**：指轻质量货物的运输。
Zero-Emission Vehicle: A clean fuel vehicle meeting even more stringent zero-emission vehicle standards.	**零排放汽车**：使用清洁燃料的汽车，达到严格的汽车尾气零排放标准。
Zone Charge: An extra fee charged for crossing a predetermined boundary.	**地段收费**：指因穿过预先确定的分界线而额外收取的费用。（APTA1）

Index of Chinese Terms